The Fair Haven by Samuel Butler

Samuel Butler was born on 4th December 1835 at the village rectory in Langar, Nottinghamshire.

His relationship with his parents, especially his father, was largely antagonistic. His education began at home and included frequent beatings, as was all too common at the time.

Under his parents' influence, he was set to follow his father into the priesthood. He was schooled at Shrewsbury and then St John's College, Cambridge, where he obtained a first in Classics in 1858.

After Cambridge he went to live in a low-income parish in London 1858–59 as preparation for his ordination into the Anglican clergy; there he discovered that baptism made no apparent difference to the morals and behaviour of his new peers. He began to question his faith. Correspondence with his father about the issue failed to set his mind at peace, inciting instead his father's wrath.

As a result, the young Butler emigrated in September 1859 to New Zealand. He was determined to change his life.

He wrote of his arrival and life as a sheep farmer on Mesopotamia Station in 'A First Year in Canterbury Settlement' (1863). After a few years he sold his farm and made a handsome profit. But the chief achievement of these years were the drafts and source material for much of his masterpiece 'Erewhon'.

Butler returned to England in 1864, settling in rooms in Clifford's Inn, near Fleet Street, where he would live for the rest of his life.

In 1872, he published his Utopian novel 'Erewhon' which made him a well-known figure.

He wrote a number of other books, including a moderately successful sequel, 'Erewhon Revisited' before his masterpiece and semi-autobiographical novel 'The Way of All Flesh' appeared after his death. Butler thought its tone of satirical attack on Victorian morality too contentious to publish during his life time and thereby shied away from further potential problems.

Samuel Butler died aged 66 on 18th June 1902 at a nursing home in St John's Wood Road, London. He was cremated at Woking Crematorium, and accounts say his ashes were either dispersed or buried in an unmarked grave.

Index of Contents

INTRODUCTION by R. A. Streatfeild

The demand for a new edition of The Fair Haven gives me an opportunity of saying a few words about the genesis of what, though not one of the most popular of Samuel Butler's books, is certainly one of the most characteristic. Few of his works, indeed, show more strikingly his brilliant powers as a controversialist and his implacable determination to get at the truth of whatever engaged his attention.

To find the germ of The Fair Haven we should probably have to go back to the year 1858, when Butler, after taking his degree at Cambridge, was preparing himself for holy orders by acting as a kind of lay curate in a London parish. Butler never took things for granted, and he felt it to be his duty to examine independently a good many points of Christian dogma which most candidates for ordination accept as matters of course. The result of his investigations was that he eventually declined to take orders at all. One of the stones upon which he then stumbled was the efficacy of infant baptism, and I have no doubt that another was the miraculous element of Christianity, which, it will be remembered, was the cause of grievous searchings of heart to Ernest Pontifex in Butler's semi-autobiographical novel, The Way of All Flesh. While Butler was in New Zealand (1859–64) he had leisure for prosecuting his Biblical studies, the result of which he published in 1865, after his return to England, in an anonymous pamphlet entitled "The Evidence for the Resurrection of Jesus Christ as given by the Four Evangelists critically examined." This pamphlet passed unnoticed; probably only a few copies were printed and it is now extremely rare. After the publication of Erewhon in 1872, Butler returned once more to theology, and made his anonymous pamphlet the basis of the far more elaborate Fair Haven, which was originally published as the posthumous work of a certain John Pickard Owen, preceded by a memoir of the deceased author by his supposed brother, William Bickersteth Owen. It is possible that the memoir was the fruit of a suggestion made by Miss Savage, an able and witty woman with whom Butler corresponded at the time. Miss Savage was so much impressed by the narrative power displayed in Erewhon that she urged Butler to write a novel, and we shall probably not be far wrong in regarding the biography of John Pickard Owen as Butler's trial trip in the art of fiction—a prelude to The Way of All Flesh, which he began in 1873.

It has often been supposed that the elaborate paraphernalia of mystification which Butler used in The Fair Haven was deliberately designed in order to hoax the public. I do not believe that this was the case. Butler, I feel convinced, provided an ironical framework for his arguments merely that he might render them more effective than they had been when plainly stated in the pamphlet of 1865. He fully expected his readers to comprehend his irony, and he anticipated that some at any rate of them would keenly resent it. Writing to Miss Savage in March, 1873 (shortly before the publication of the book), he said: "I should hope that attacks on The Fair Haven will give me an opportunity of excusing myself, and if so I shall endeavour that the excuse may be worse than the fault it is intended to excuse." A few days later he referred to the difficulties that he had encountered in getting the book accepted by a publisher: "—

were frightened and even considered the scheme of the book unjustifiable. — urged me, as politely as he could, not to do it, and evidently thinks I shall get myself into disgrace even among freethinkers. It's all nonsense. I dare say I shall get into a row—at least I hope I shall." Evidently there is here no anticipation of The Fair Haven being misunderstood. Misunderstood, however, it was, not only by reviewers, some of whom greeted it solemnly as a defence of orthodoxy, but by divines of high standing, such as the late Canon Ainger, who sent it to a friend whom he wished to convert. This was more than Butler could resist, and he hastened to issue a second edition bearing his name and accompanied by a preface in which the deceived elect were held up to ridicule.

Butler used to maintain that The Fair Haven did his reputation no harm. Writing in 1901, he said:

"The Fair Haven got me into no social disgrace that I have ever been able to discover. I might attack Christianity as much as I chose and nobody cared one straw; but when I attacked Darwin it was a different matter. For many years Evolution, Old and New, and Unconscious Memory made a shipwreck of my literary prospects. I am only now beginning to emerge from the literary and social injury which those two perfectly righteous books inflicted on me. I dare say they abound with small faults of taste, but I rejoice in having written both of them."

Very likely Butler was right as to the social side of the question, but I am convinced that The Fair Haven did him grave harm in the literary world. Reviewers fought shy of him for the rest of his life. They had been taken in once, and they took very good care that they should not be taken in again. The word went forth that Butler was not to be taken seriously, whatever he wrote, and the results of the decree were apparent in the conspiracy of silence that greeted not only his books on evolution, but his Homeric works, his writings on art, and his edition of Shakespeare's sonnets. Now that he has passed beyond controversies and mystifications, and now that his other works are appreciated at their true value, it is not too much to hope that tardy justice will be accorded also to The Fair Haven. It is true that the subject is no longer the burning question that it was forty years ago. In the early seventies theological polemics were fashionable. Books like Seeley's Ecce Homo and Matthew Arnold's Literature and Dogma were eagerly devoured by readers of all classes. Nowadays we take but a languid interest in the problems that disturbed our grandfathers, and most of us have settled down into what Disraeli described as the religion of all sensible men, which no sensible man ever talks about. There is, however, in The Fair Haven a good deal more than theological controversy, and our Laodicean age will appreciate Butler's humour and irony if it cares little for his polemics. The Fair Haven scandalised a good many people when it first appeared, but I am not afraid of its scandalising anybody now. I should be sorry, nevertheless, if it gave any reader a false impression of Butler's Christianity, and I think I cannot do better than conclude with a passage from one of his essays which represents his attitude to religion perhaps more faithfully than anything in The Fair Haven: "What, after all, is the essence of Christianity? What is the kernel of the nut? Surely common sense and cheerfulness, with unflinching opposition to the charlatanisms and Pharisaisms of a man's own times. The essence of Christianity lies neither in dogma, nor yet in abnormally holy life, but in faith in an unseen world, in doing one's duty, in speaking the truth, in finding the true life rather in others than in oneself, and in the certain hope that he who loses his life on these behalfs finds more than he has lost. What can Agnosticism do against such Christianity as this? I should be shocked if anything I had ever written or shall ever write should seem to make light of these things."

R. A. STREATFEILD.
August, 1913.

The occasion of a Second Edition of The Fair Haven enables me to thank the public and my critics for the favourable reception which has been accorded to the First Edition. I had feared that the freedom with which I had exposed certain untenable positions taken by Defenders of Christianity might have given offence to some reviewers, but no complaint has reached me from any quarter on the score of my not having put the best possible case for the evidence in favour of the miraculous element in Christ's teaching—nor can I believe that I should have failed to hear of it, if my book had been open to exception on this ground.

An apology is perhaps due for the adoption of a pseudonym, and even more so for the creation of two such characters as JOHN PICKARD OWEN and his brother. Why could I not, it may be asked, have said all that I had to say in my own proper person?

Are there not real ills of life enough already? Is there not a "lo here!" from this school with its gushing "earnestness," it distinctions without differences, its gnat strainings and camel swallowings, its pretence of grappling with a question while resolutely bent upon shirking it, its dust throwing and mystification, its concealment of its own ineffable insincerity under an air of ineffable candour? Is there not a "lo there!" from that other school with its bituminous atmosphere of exclusiveness and self-laudatory dilettanteism? Is there not enough actual exposition of boredom come over us from many quarters without drawing for new bores upon the imagination? It is true I gave a single drop of comfort. JOHN PICKARD OWEN was dead. But his having ceased to exist (to use the impious phraseology of the present day) did not cancel the fact of his having once existed. That he should have ever been born gave proof of potentialities in Nature which could not be regarded lightly. What hybrids might not be in store for us next? Moreover, though JOHN PICKARD was dead, WILLIAM BICKERSTETH was still living, and might at any moment rekindle his burning and shining lamp of persistent self-satisfaction. Even though the OWENS had actually existed, should not their existence have been ignored as a disgrace to Nature? Who then could be justified in creating them when they did not exist?

I am afraid I must offer an apology rather than an excuse. The fact is that I was in a very awkward position. My previous work, Erewhon, had failed to give satisfaction to certain ultra-orthodox Christians, who imagined that they could detect an analogy between the English Church and the Erewhonian Musical Banks. It is inconceivable how they can have got hold of this idea; but I was given to understand that I should find it far from easy to dispossess them of the notion that something in the way of satire had been intended. There were other parts of the book which had also been excepted to, and altogether I had reason to believe that if I defended Christianity in my own name I should not find Erewhon any addition to the weight which my remarks might otherwise carry. If I had been suspected of satire once, I might be suspected again with no greater reason. Instead of calmly reviewing the arguments which I adduced, The Rock might have raised a cry of non tali auxilio. It must always be remembered that besides the legitimate investors in Christian stocks, if so homely a metaphor may be pardoned, there are unscrupulous persons whose profession it is to be bulls, bears, stags, and I know not what other creatures of the various Christian markets. It is all nonsense about hawks not picking out each other's eyes—there is nothing they like better. I feared The Guardian, The Record, The John Bull, etc., lest they should suggest that from a bear I now turned bull with a view to an eventual bishopric. Such insinuations would have impaired the value of The Fair Haven as an anchorage for well-meaning people. I therefore resolved to obey the injunction of the Gentile Apostle and avoid all appearance of evil, by dissociating myself from the author of Erewhon as completely as possible. At the moment of my resolution JOHN PICKARD OWEN came to my assistance; I felt that he was the sort of man I wanted, but that he was hardly sufficient in himself. I therefore summoned his brother. The pair have served their purpose; a year nowadays produces great changes in men's thoughts concerning Christianity, and the little matter of Erewhon having quite blown over I feel that I may safely appear in my true colours as the champion of orthodoxy, discard the OWENS as other than mouthpieces, and relieve the public from uneasiness as to any further writings from the pen of the surviving brother.

Nevertheless I am bound to own that, in spite of a generally favourable opinion, my critics have not been unanimous in their interpretation of The Fair Haven. Thus, The Rock (April 25, 1873, and May 9, 1873), says that the work is "an extraordinary one, whether regarded as a biographical record or a theological treatise. Indeed the importance of the volume compels us to depart from our custom of reviewing with brevity works entrusted to us, and we shall in two consecutive numbers of The Rock lay before its readers what appear to us to be the merits and demerits of this posthumous production."

"His exhibition of the certain proofs furnished of the Resurrection of our Lord is certainly masterly and convincing."

"To the sincerely inquiring doubter, the striking way in which the truth of the Resurrection is exhibited must be most beneficial, but such a character we are compelled to believe is rare among those of the schools of neology."

"Mr. OWEN'S exposition and refutation of the hallucination and mythical theories of Strauss and his followers is most admirable, and all should read it who desire to know exactly what excuses men make for their incredulity. The work also contains many beautiful passages on the discomfort of unbelief, and the holy pleasure of a settled faith, which cannot fail to benefit the reader."

On the other hand, in spite of all my precautions, the same misfortune which overtook Erewhon has also come upon The Fair Haven. It has been suspected of a satirical purpose. The author of a pamphlet entitled Jesus versus Christianity says:—

"The Fair Haven is an ironical defence of orthodoxy at the expense of the whole mass of Church tenet and dogma, the character of Christ only excepted. Such at least is our reading of it, though critics of the Rock and Record order have accepted the book as a serious defence of Christianity, and proclaimed it as a most valuable contribution in aid of the faith. Affecting an orthodox standpoint it most bitterly reproaches all previous apologists for the lack of candour with which they have ignored or explained away insuperable difficulties and attached undue value to coincidences real or imagined. One and all they have, the author declares, been at best, but zealous 'liars for God,' or what to them was more than God, their own religious system. This must go on no longer. We, as Christians having a sound cause, need not fear to let the truth be known. He proceeds accordingly to set forth the truth as he finds it in the New Testament; and in a masterly analysis of the account of the Resurrection, which he selects as the principal crucial miracle, involving all other miracles, he shows how slender is the foundation on which the whole fabric of supernatural theology has been reared."

"As told by our author the whole affords an exquisite example of the natural growth of a legend."

"If the reader can once fully grasp the intention of the style, and its affectation of the tone of indignant orthodoxy, and perceive also how utterly destructive are its 'candid admissions' to the whole fabric of supernaturalism, he will enjoy a rare treat. It is not however for the purpose of recommending what we at least regard as a piece of exquisite humour, that we call attention to The Fair Haven, but &c. &c."

This is very dreadful; but what can one do?

Again, The Scotsman speaks of the writer as being "throughout in downright almost pathetic earnestness." While The National Reformer seems to be in doubt whether the book is a covert attack upon Christianity or a serious defence of it, but declares that both orthodox and unorthodox will find matter requiring thought and answer.

I am not responsible for the interpretations of my readers. It is only natural that the same work should present a very different aspect according as it is approached from one side or the other. There is only one way out of it—that the reader should kindly interpret according to his own fancies. If he will do this the book is sure to please him. I have done the best I can for all parties, and feel justified in appealing to the existence of the widely conflicting opinions which I have quoted, as a proof that the balance has been evenly held, and that I was justified in calling the book a defence—both as against impugners and defenders.

S. BUTLER.
Oct. 8, 1873.

MEMOIR OF THE LATE JOHN PICKARD OWEN

CHAPTER I

The subject of this Memoir, and Author of the work which follows it, was born in Goodge Street, Tottenham Court Road, London, on the 5th of February, 1832. He was my elder brother by about eighteen months. Our father and mother had once been rich, but through a succession of unavoidable misfortunes they were left with but a very moderate income when my brother and myself were about three and four years old. My father died some five or six years afterwards, and we only recollected him as a singularly gentle and humorous playmate who doted upon us both and never spoke unkindly. The charm of such a recollection can never be dispelled; both my brother and myself returned his love with interest, and cherished his memory with the most affectionate regret, from the day on which he left us till the time came that the one of us was again to see him face to face. So sweet and winning was his nature that his slightest wish was our law—and whenever we pleased him, no matter how little, he never failed to thank us as though we had done him a service which we should have had a perfect right to withhold. How proud were we upon any of these occasions, and how we courted the opportunity of being thanked! He did indeed well know the art of becoming idolised by his children, and dearly did he prize the results of his own proficiency; yet truly there was no art about it; all arose spontaneously from the wellspring of a sympathetic nature which knew how to feel as others felt, whether old or young, rich or poor, wise or foolish. On one point alone did he neglect us—I refer to our religious education. On all other matters he was the kindest and most careful teacher in the world. Love and gratitude be to his memory!

My mother loved us no less ardently than my father, but she was of a quicker temper, and less adept at conciliating affection. She must have been exceedingly handsome when she was young, and was still comely when we first remembered her; she was also highly accomplished, but she felt my father's loss of fortune more keenly than my father himself, and it preyed upon her mind, though rather for our sake than for her own. Had we not known my father we should have loved her better than any one in the world, but affection goes by comparison, and my father spoiled us for any one but himself; indeed, in after life, I remember my mother's telling me, with many tears, how jealous she had often been of the love we bore him, and how mean she had thought it of him to entrust all scolding or repression to her, so that he might have more than his due share of our affection. Not that I believe my father did this consciously; still, he so greatly hated scolding that I dare say we might often have got off scot free when we really deserved reproof had not my mother undertaken the onus of scolding us herself. We therefore naturally feared her more than my father, and fearing more we loved less. For as love casteth out fear, so fear love.

This must have been hard to bear, and my mother scarcely knew the way to bear it. She tried to upbraid us, in little ways, into loving her as much as my father; the more she tried this, the less we could succeed in doing it; and so on and so on in a fashion which need not be detailed. Not but what we really loved her deeply, while her affection for us was unsurpassable still, we loved her less than we loved my father, and this was the grievance.

My father entrusted our religious education entirely to my mother. He was himself, I am assured, of a deeply religious turn of mind, and a thoroughly consistent member of the Church of England; but he conceived, and perhaps rightly, that it is the mother who should first teach her children to lift their hands in prayer, and impart to them a knowledge of the One in whom we live and move and have our being. My mother accepted the task gladly, for in spite of a certain narrowness of view—the natural but deplorable result of her earlier surroundings—she was one of the most truly pious women whom I have ever known; unfortunately for herself and us she had been trained in the lowest school of Evangelical literalism—a school which in after life both my brother and myself came to regard as the main obstacle to the complete overthrow of unbelief; we therefore looked upon it with something stronger than aversion, and for my own part I still deem it perhaps the most insidious enemy which the cause of Christ has ever encountered. But of this more hereafter.

My mother, as I said, threw her whole soul into the work of our religious education. Whatever she believed she believed literally, and, if I may say so, with a harshness of realisation which left very little scope for imagination or mystery. Her plans of Heaven and solutions of life's enigmas were direct and forcible, but they could only be reconciled with certain obvious facts—such as the omnipotence and all-goodness of God—by leaving many things absolutely out of sight. And this my mother succeeded effactually in doing. She never doubted that her opinions comprised the truth, the whole truth, and nothing but the truth; she therefore made haste to sow the good seed in our tender minds, and so far succeeded that when my brother was four years old he could repeat the Apostles' Creed, the General Confession, and the Lord's Prayer without a blunder. My mother made herself believe that he delighted in them; but, alas! it was far otherwise; for, strange as it may appear concerning one whose later life was a continual prayer, in childhood he detested nothing so much as being made to pray and to learn his Catechism. In this I am sorry to say we were both heartily of a mind. As for Sunday, the less said the better.

I have already hinted (but as a warning to other parents I had better, perhaps, express myself more plainly), that this aversion was probably the result of my mother's undue eagerness to reap an artificial fruit of lip service, which could have little meaning to the heart of one so young. I believe that the severe check which the natural growth of faith experienced in my brother's case was due almost entirely to this cause, and to the school of literalism in which he had been trained; but, however this may be, we both of us hated being made to say our prayers—morning and evening it was our one bugbear, and we would avoid it, as indeed children generally will, by every artifice which we could employ. Thus we were in the habit of feigning to be asleep shortly before prayer time, and would gratefully hear my father tell my mother that it was a shame to wake us; whereon he would carry us up to bed in a state apparently of the profoundest slumber when we were really wide awake and in great fear of detection. For we knew how to pretend to be asleep, but we did not know how we ought to wake again; there was nothing for it therefore when we were once committed, but to go on sleeping till we were fairly undressed and put to bed, and could wake up safely in the dark. But deceit is never long successful, and we were at last ignominiously exposed.

It happened one evening that my mother suspected my brother John, and tried to open his little hands which were lying clasped in front of him. Now my brother was as yet very crude and inconsistent in his theories concerning sleep, and had no conception of what a real sleeper would do under these circumstances. Fear deprived him of his powers of reflection, and he thus unfortunately concluded that because sleepers, so far as he had observed them, were always motionless, therefore, they must be quite rigid and incapable of motion, and indeed that any movement, under any circumstances (for from his earliest childhood he liked to carry his theories to their legitimate conclusion), would be physically impossible for one who was really sleeping; forgetful, oh! unhappy one, of the flexibility of his own body on being carried upstairs, and, more unhappy still, ignorant of the art of waking. He, therefore, clenched his fingers harder and harder as he felt my mother trying to unfold them while his head hung listless, and his eyes were closed I as though he were sleeping sweetly. It is needless to detail the agony of shame that followed. My mother begged my father to box his ears, which my father flatly refused to do. Then she boxed them herself, and there followed a scene and a day or two of disgrace for both of us.

Shortly after this there happened another misadventure. A lady came to stay with my mother, and was to sleep in a bed that had been brought into our nursery, for my father's fortunes had already failed, and we were living in a humble way. We were still but four and five years old, so the arrangement was not unnatural, and it was assumed that we should be asleep before the lady went to bed, and be downstairs before she would get up in the morning. But the arrival of this lady and her being put to sleep in the nursery were great events to us in those days, and being particularly wanted to go to sleep, we of course sat up in bed talking and keeping ourselves awake till she should come upstairs. Perhaps we had fancied that she would give us something, but if so we were disappointed. However, whether this was the case or not, we were wide awake when our visitor came to bed, and having no particular object to gain, we made no pretence of sleeping. The lady kissed us both, told us to lie still and go to sleep like good children, and then began doing her hair.

I remember that this was the occasion on which my brother discovered a good many things in connection with the fair sex which had hitherto been beyond his ken; more especially that the mass of petticoats and clothes which envelop the female form were not, as he expressed it to me, "all solid woman," but that women were not in reality more substantially built than men, and had legs as much as he had, a fact which he had never yet realised. On this he for a long time considered them as impostors, who had wronged him by leading him to suppose that they had far more "body in them" (so he said), than he now found they had. This was a sort of thing which he regarded with stern moral reprobation. If he had been old enough to have a solicitor I believe he would have put the matter into his hands, as well as certain other things which had lately troubled him. For but recently my mother had bought a fowl, and he had seen it plucked, and the inside taken out; his irritation had been extreme on discovering that fowls were not all solid flesh, but that their insides—and these formed, as it appeared to him, an enormous percentage of the bird—were perfectly useless. He was now beginning to understand that sheep and cows were also hollow as far as good meat was concerned; the flesh they had was only a mouthful in comparison with what they ought to have considering their apparent bulk—insignificant, mere skin and bone covering a cavern. What right had they, or anything else, to assert themselves as so big, and prove so empty? And now this discovery of woman's falsehood was quite too much for him. The world itself was hollow, made up of shams and delusions, full of sound and fury signifying nothing.

Truly a prosaic young gentleman enough. Everything with him was to be exactly in all its parts what it appeared on the face of it, and everything was to go on doing exactly what it had been doing hitherto. If a thing looked solid, it was to be very solid; if hollow, very hollow; nothing was to be half and half, and nothing was to change unless he had himself already become accustomed to its times and manners of changing; there were to be no exceptions and no contradictions; all things were to be perfectly consistent, and all premises to be carried with extremest rigour to their legitimate conclusions. Heaven was to be very neat (for he was always tidy himself), and free from sudden shocks to the nervous system, such as those caused by dogs barking at him, or cows driven in the streets. God was to resemble my father, and the Holy Spirit to bear some sort of indistinct analogy to my mother.

Such were the ideal theories of his childhood—unconsciously formed, but very firmly believed in. As he grew up he made such modifications as were forced upon him by enlarged perceptions, but every modification was an effort to him, in spite of a continual and successful resistance to what he recognised as his initial mental defect.

I may perhaps be allowed to say here, in reference to a remark in the preceding paragraph, that both my brother and myself used to notice it as an almost invariable rule that children's earliest ideas of God are modelled upon the character of their father—if they have one. Should the father be kind, considerate, full of the warmest love, fond of showing it, and reserved only about his displeasure, the child having learned to look upon God as His Heavenly Father through the Lord's Prayer and our Church Services, will feel towards God as he does towards his own father; this conception will stick to a man for years and years after he has attained manhood—probably it will never leave him. For all children love their fathers and mothers, if these last will only let them; it is not a little unkindness that will kill so hardy a plant as the love of a child for its parents. Nature has allowed ample margin for many blunders, provided there be a genuine desire on the parent's part to make the child feel that he is loved, and that his natural feelings are respected. This is all the religious education which a child should have. As he grows older he will then turn naturally to the waters of life, and thirst after them of his own accord by reason of the spiritual refreshment which they, and they only, can afford. Otherwise he will shrink from them, on account of his recollection of the way in which he was led down to drink against his will, and perhaps with harshness, when all the analogies with which he was acquainted pointed in the direction of their being unpleasant and unwholesome. So soul-satisfying is family affection to a child, that he who has once enjoyed it cannot bear to be deprived of the hope that he is possessed in Heaven of a parent who is like his earthly father—of a friend and counsellor who will never, never fail him. There is no such religious nor moral education as kindly genial treatment and a good example; all else may then be let alone till the child is old enough to feel the want of it. It is true that the seed will thus be sown late, but in what a soil! On the other hand, if a man has found his earthly father harsh and uncongenial, his conception of his Heavenly Parent will be painful. He will begin by seeing God as an exaggerated likeness of his father. He will therefore shrink from Him. The rottenness of stillborn love in the heart of a child poisons the blood of the soul, and hence, later, crime.

To return, however, to the lady. When she had put on her night-gown, she knelt down by her bedside and, to our consternation, began to say her prayers. This was a cruel blow to both of us; we had always been under the impression that grownup people were not made to say their prayers, and the idea of any one saying them of his or her own accord had never occurred to us as possible. Of course the lady would not say her prayers if she were not obliged; and yet she did say them; therefore she must be obliged to say them; therefore we should be obliged to say them, and this was a very great disappointment. Awe-struck and open-mouthed we listened while the lady prayed in sonorous accents, for many things which I do not now remember, and finally for my father and mother and for both of us—shortly afterwards she rose, blew out the light and got into bed. Every word that she said had confirmed our worst apprehensions; it was just what we had been taught to say ourselves.

Next morning we compared notes and drew the most painful inferences; but in the course of the day our spirits rallied. We agreed that there were many mysteries in connection with life and things which it was high time to unravel, and that an opportunity was now afforded us which might not readily occur again. All we had to do was to be true to ourselves and equal to the occasion. We laid our plans with great astuteness. We would be fast asleep when the lady came up to bed, but our heads should be turned in the direction of her bed, and covered with clothes, all but a single peep-hole. My brother, as the eldest, had clearly a right to be nearest the lady, but I could see very well, and could depend on his reporting faithfully whatever should escape me.

There was no chance of her giving us anything—if she had meant to do so she would have done it sooner; she might, indeed, consider the moment of her departure as the most auspicious for this purpose, but then she was not going yet, and the interval was at our own disposal. We spent the afternoon in trying to learn to snore, but we were not certain about it, and in the end regretfully concluded that as snoring was not de rigueur we had better dispense with it.

We were put to bed; the light was taken away; we were told to go to sleep, and promised faithfully that we would do so; the tongue indeed swore, but the mind was unsworn. It was agreed that we should keep pinching one another to prevent our going to sleep. We did so at frequent intervals; at last our patience was rewarded with the heavy creak, as of a stout elderly lady labouring up the stairs, and presently our victim entered.

To cut a long story short, the lady on satisfying herself that we were asleep, never said her prayers at all; during the remainder of her visit whenever she found us awake she always said them, but when she thought we were asleep, she never prayed. It is needless to add that we had the matter out with her before she left, and that the consequences were unpleasant for all parties; they added to the troubles in which we were already involved as to our prayers, and were indirectly among the earliest causes which led my brother to look with scepticism upon religion.

For a while, however, all went on as though nothing had happened. An effect of distrust, indeed, remained after the cause had been forgotten, but my brother was still too young to oppose anything that my mother told him, and to all outward appearance he grew in grace no less rapidly than in stature.

For years we led a quiet and eventless life, broken only by the one great sorrow of our father's death. Shortly after this we were sent to a day school in Bloomsbury. We were neither of us very happy there, but my brother, who always took kindly to his books, picked up a fair knowledge of Latin and Greek; he also learned to draw, and to exercise himself a little in English composition. When I was about fourteen my mother capitalised a part of her income and started me off to America, where she had friends who could give me a helping hand; by their kindness I was enabled, after an absence of twenty years, to return with a handsome income, but not, alas, before the death of my mother.

Up to the time of my departure my mother continued to read the Bible with us and explain it. She had become deeply impressed with the millenarian fervour which laid hold of so many some twenty-five or thirty years ago. The Apocalypse was perhaps her favourite book in the Bible, and she was imbued with the fullest conviction that all the threatened horrors with which it teems were upon the eve of their accomplishment. The year eighteen hundred and forty-eight was to be (as indeed it was) a time of general bloodshed and confusion, while in eighteen hundred and sixty-six, should it please God to spare her, her eyes would be gladdened by the visible descent of the Son of Man with a shout, with the voice of the Archangel, with the trump of God; and the dead in Christ should rise first; then she, as one of them that were alive, would be caught up with other saints into the air, and would possibly receive while rising some distinguishing token of confidence and approbation which should fall with due impressiveness upon the surrounding multitude; then would come the consummation of all things, and she would be ever with the Lord. She died peaceably in her bed before she could know that a commercial panic was the nearest approach to the fulfilment of prophecy which the year eighteen hundred and sixty-six brought forth.

These opinions of my mother's were positively disastrous—injuring her naturally healthy and vigorous mind by leading her to indulge in all manner of dreamy and fanciful interpretations of Scripture, which any but the most narrow literalist would feel at once to be untenable. Thus several times she expressed to us her conviction that my brother and myself were to be the two witnesses mentioned in the eleventh chapter of the Book of Revelation, and dilated upon the gratification she should experience upon finding that we had indeed been reserved for a position of such distinction. We were as yet mere children, and naturally took all for granted that our mother told us; we therefore made a careful examination of the passage which threw light upon our future; but on finding that the prospect was gloomy and full of bloodshed we protested against the honours which were intended for us, more especially when we reflected that the mother of the two witnesses was not menaced in Scripture with any particular discomfort. If we were to be martyrs, my mother ought to wish to be a martyr too, whereas nothing was farther from her intention. Her notion clearly was that we were to be massacred somewhere in the streets of London, in consequence of the anti-Christian machinations of the Pope; that after lying about unburied for three days and a half we were to come to life again; and, finally, that we should conspicuously ascend to heaven, in front, perhaps, of the Foundling Hospital.

She was not herself indeed to share either our martyrdom or our glorification, but was to survive us many years on earth, living in an odour of great sanctity and reflected splendour, as the central and most august figure in a select society. She would perhaps be able indirectly, through her sons' influence with the Almighty, to have a voice in most of the arrangements both of this world and of the next. If all this were to come true (and things seemed very like it), those friends who had neglected us in our adversity would not find it too easy to be restored to favour, however greatly they might desire it—that is to say, they would not have found it too easy in the case of one less magnanimous and spiritually-minded than herself. My mother said but little of the above directly, but the fragments which occasionally escaped her were pregnant, and on looking back it is easy to perceive that she must have been building one of the most stupendous aerial fabrics that have ever been reared.

I have given the above in its more amusing aspect, and am half afraid that I may appear to be making a jest of weakness on the part of one of the most devotedly unselfish mothers who have ever existed. But one can love while smiling, and the very wildness of my mother's dream serves to show how entirely her whole soul was occupied with the things which are above. To her, religion was all in all; the earth was but a place of pilgrimage—only so far important as it was a possible road to heaven. She impressed this upon both of us by every word and action—instant in season and out of season, so that she might fill us more deeply with a sense of God. But the inevitable consequences happened; my mother had aimed too high and had overshot her mark. The influence indeed of her guileless and unworldly nature remained impressed upon my brother even during the time of his extremest unbelief (perhaps his ultimate safety is in the main referable to this cause, and to the happy memories of my father, which had predisposed him to love God), but my mother had insisted on the most minute verbal accuracy of every part of the Bible; she had also dwelt upon the duty of independent research, and on the necessity of giving up everything rather than assent to things which our conscience did not assent to. No one could have more effectually taught us to try to think the truth, and we had taken her at her word because our hearts told us that she was right. But she required three incompatible things. When my brother grew older he came to feel that independent and unflinching examination, with a determination to abide by the results, would lead him to reject the point which to my mother was more important than any other—I mean the absolute accuracy of the Gospel records. My mother was inexpressibly shocked at hearing my brother doubt the authenticity of the Epistle to the Hebrews; and then, as it appeared to him, she tried to make him violate the duties of examination and candour which he had learnt too thoroughly to unlearn. Thereon came pain and an estrangement which was none the less profound for being mutually concealed.

This estrangement was the gradual work of some five or six years, during which my brother was between eleven and seventeen years old. At seventeen, I am told that he was remarkably well informed and clever. His manners were, like my father's, singularly genial, and his appearance very prepossessing. He had as yet no doubt concerning the soundness of any fundamental Christian doctrine, but his mind was too active to allow of his being contented with my mother's child-like faith. There were points on which he did not indeed doubt, but which it would none the less be interesting to consider; such for example as the perfectibility of the regenerate Christian, and the meaning of the mysterious central chapters of the Epistle to the Romans. He was engaged in these researches though still only a boy, when an event occurred which gave the first real shock to his faith.

He was accustomed to teach in a school for the poorest children every Sunday afternoon, a task for which his patience and good temper well fitted him. On one occasion, however, while he was explaining the effect of baptism to one of his favourite pupils, he discovered to his great surprise that the boy had never been baptised. He pushed his inquiries further, and found that out of the fifteen boys in his class only five had been baptised, and, not only so, but that no difference in disposition or conduct could be discovered between the regenerate boys and the unregenerate. The good and bad boys were distributed in proportions equal to the respective numbers of the baptised and unbaptised. In spite of a certain impetuosity of natural character, he was also of a matter-of-fact and experimental turn of mind; he therefore went through the whole school, which numbered about a hundred boys, and found out who had been baptised and who had not. The same results appeared. The majority had not been baptised; yet the good and bad dispositions were so distributed as to preclude all possibility of maintaining that the baptised boys were better than the unbaptised.

The reader may smile at the idea of any one's faith being troubled by a fact of which the explanation is so obvious, but in truth my brother was seriously and painfully shocked. The teacher to whom he applied for a solution of the difficulty was not a man of any real power, and reported my brother to the rector for having disturbed the school by his inquiries. The rector was old and self-opinionated; the difficulty, indeed, was plainly as new to him as it had been to my brother, but instead of saying so at once, and referring to any recognised theological authority, he tried to put him off with words which seemed intended to silence him rather than to satisfy him; finally he lost his temper, and my brother fell under suspicion of unorthodoxy.

This kind of treatment might answer with some people, but not with my brother. He alludes to it resentfully in the introductory chapter of his book. He became suspicious that a preconceived opinion was being defended at the expense of honest scrutiny, and was thus driven upon his own unaided investigation. The result may be guessed: he began to go astray, and strayed further and further. The children of God, he reasoned, the members of Christ and inheritors of the kingdom of Heaven, were no more spiritually minded than the children of the world and the devil. Was then the grace of God a gift which left no trace whatever upon those who were possessed of it—a thing the presence or absence of which might be ascertained by consulting the parish registry, but was not discernible in conduct? The grace of man was more clearly perceptible than this. Assuredly there must be a screw loose somewhere, which, for aught he knew, might be jeopardising the salvation of all Christendom. Where then was this loose screw to be found?

He concluded after some months of reflection that the mischief was caused by the system of sponsors and by infant baptism. He therefore, to my mother's inexpressible grief, joined the Baptists and was immersed in a pond near Dorking. With the Baptists he remained quiet about three months, and then began to quarrel with his instructors as to their doctrine of predestination. Shortly afterwards he came accidentally upon a fascinating stranger who was no less struck with my brother than my brother with him, and this gentleman, who turned out to be a Roman Catholic missionary, landed him in the Church of Rome, where he felt sure that he had now found rest for his soul. But here, too, he was mistaken; after about two years he rebelled against the stifling of all free inquiry; on this rebellion the flood-gates of scepticism were opened, and he was soon battling with unbelief. He then fell in with one who was a pure Deist, and was shorn of every shred of dogma which he had ever held, except a belief in the personality and providence of the Creator.

On reviewing his letters written to me about this time, I am painfully struck with the manner in which they show that all these pitiable vagaries were to be traced to a single cause—a cause which still exists to the misleading of hundreds of thousands, and which, I fear, seems likely to continue in full force for many a year to come—I mean, to a false system of training which teaches people to regard Christianity as a thing one and indivisible, to be accepted entirely in the strictest reading of the letter, or to be rejected as absolutely untrue. The fact is, that all permanent truth is as one of those coal measures, a seam of which lies near the surface, and even crops up above the ground, but which is generally of an inferior quality and soon worked out; beneath it there comes a layer of sand and clay, and then at last the true seam of precious quality and in virtually inexhaustible supply. The truth which is on the surface is rarely the whole truth. It is seldom until this has been worked out and done with—as in the case of the apparent flatness of the earth—that unchangeable truth is discovered. It is the glory of the Lord to conceal a matter: it is the glory of the king to find it out. If my brother, from whom I have taken the above illustration, had had some judicious and wide-minded friend to correct and supplement the mainly admirable principles which had been instilled into him by my mother, he would have been saved years of spiritual wandering; but, as it was, he fell in with one after another, each in his own way as literal and unspiritual as the other—each impressed with one aspect of religious truth, and with one only. In the end he became perhaps the widest-minded and most original thinker whom I have ever met; but no one from his early manhood could have augured this result; on the contrary, he shewed every sign of being likely to develop into one of those who can never see more than one side of a question at a time, in spite of their seeing that side with singular clearness of mental vision. In after life, he often met with mere lads who seemed to him to be years and years in advance of what he had been at their age, and would say, smiling, "With a great sum obtained I this freedom; but thou wast free-born."

Yet when one comes to think of it, a late development and laborious growth are generally more fruitful than those which are over-early luxuriant. Drawing an illustration from the art of painting, with which he was well acquainted, my brother used to say that all the greatest painters had begun with a hard and precise manner from which they had only broken after several years of effort; and that in like manner all the early schools were founded upon definiteness of outline to the exclusion of truth of effect. This may be true; but in my brother's case there was something even more unpromising than this; there was a commonness, so to speak, of mental execution, from which no one could have foreseen his after-emancipation. Yet in the course of time he was indeed emancipated to the very uttermost, while his bonds will, I firmly trust, be found to have been of inestimable service to the whole human race.

For although it was so many years before he was enabled to see the Christian scheme as a whole, or even to conceive the idea that there was any whole at all, other than each one of the stages of opinion through which he was at the time passing; yet when the idea was at length presented to him by one whom I must not name, the discarded fragments of his faith assumed shape, and formed themselves into a consistently organised scheme. Then became apparent the value of his knowledge of the details of so many different sides of Christian verity. Buried in the details, he had hitherto ignored the fact that they were only the unessential developments of certain component parts. Awakening to the perception of the whole after an intimate acquaintance with the details, he was able to realise the position and meaning of all that he had hitherto experienced in a way which has been vouchsafed to few, if any others.

Thus he became truly a broad Churchman. Not broad in the ordinary and ill-considered use of the term (for the broad Churchman is as little able to sympathise with Romanists, extreme High Churchmen and Dissenters, as these are with himself—he is only one of a sect which is called by the name broad, though it is no broader than its own base), but in the true sense of being able to believe in the naturalness, legitimacy, and truth quâ Christianity even of those doctrines which seem to stand most widely and irreconcilably asunder.

CHAPTER II

But it was impossible that a mind of such activity should have gone over so much ground, and yet in the end returned to the same position as that from which it started.

So far was this from being the case that the Christianity of his maturer life would be considered dangerously heterodox by those who belong to any of the more definite or precise schools of theological thought. He was as one who has made the circuit of a mountain, and yet been ascending during the whole time of his doing so: such a person finds himself upon the same side as at first, but upon a greatly higher level. The peaks which had seemed the most important when he was in the valley were now dwarfed to their true proportions by colossal cloud-capped masses whose very existence could not have been suspected from beneath: and again, other points which had seemed among the lowest turned out to be the very highest of all—as the Finster-Aarhorn, which hides itself away in the centre of the Bernese Alps, is never seen to be the greatest till one is high and far off.

Thus he felt no sort of fear or repugnance in admitting that the New Testament writings, as we now have them, are not by any means accurate records of the events which they profess to chronicle. This, which few English Churchmen would be prepared to admit, was to him so much of an axiom that he despaired of seeing any sound theological structure raised until it was universally recognised.

And here he would probably meet with sympathy from the more advanced thinkers within the body of the Church, but so far as I know, he stood alone as recognising the wisdom of the Divine counsels in having ordained the wide and apparently irreconcilable divergencies of doctrine and character which we find assigned to Christ in the Gospels, and as finding his faith confirmed, not by the supposition that both the portraits drawn of Christ are objectively true, but that both are objectively inaccurate, and that the Almighty intended they should be inaccurate, inasmuch as the true spiritual conception in the mind of man could be indirectly more certainly engendered by a strife, a warring, a clashing, so to speak, of versions, all of them distorting slightly some one or other of the features of the original, than directly by the most absolutely correct impression which human language could convey. Even the most perfect human speech, as has been often pointed out, is a very gross and imperfect vehicle of thought. I remember once hearing him say that it was not till he was nearly thirty that he discovered "what thick and sticky fluids were air and water," how crass and dull in comparison with other more subtle fluids; he added that speech had no less deceived him, seeming, as it did, to be such a perfect messenger of thought, and being after all nothing but a shuffler and a loiterer.

With most men the Gospels are true in spite of their discrepancies and inconsistencies; with him Christianity, as distinguished from a bare belief in the objectively historical character of each part of the Gospels, was true because of these very discrepancies; as his conceptions of the Divine manner of working became wider, the very forces which had at one time shaken his faith to its foundations established it anew upon a firmer and broader base. He was gradually led to feel that the ideal presented by the life and death of our Saviour could never have been accepted by Jews at all, if its whole purport had been made intelligible during the Redeemer's life-time; that in order to insure its acceptance by a nucleus of followers it must have been endowed with a more local aspect than it was intended afterwards to wear; yet that, for the sake of its subsequent universal value, the destruction of that local complexion was indispensable; that the corruptions inseparable from vivâ voce communication and imperfect education were the means adopted by the Creator to blur the details of the ideal, and give it that breadth which could not be otherwise obtainable—and that thus the value of the ideal was indefinitely enhanced, and designedly enhanced, alike by the waste of time and by its incrustations; that all ideals gain by a certain amount of vagueness, which allows the beholder to fill in the details according to his own spiritual needs, and that no ideal can be truly universal and permanent unless it have an elasticity which will allow of this process in the minds of those who contemplate it; that it cannot become thus elastic unless by the loss of no inconsiderable amount of detail, and that thus the half, as Dr. Arnold used to say, "becomes greater than the whole," the sketch more preciously suggestive than the photograph. Hence far from deploring the fragmentary, confused, and contradictory condition of the Gospel records, he saw in this condition the means whereby alone the human mind could have been enabled to conceive—not the precise nature of Christ—but the highest ideal of which each individual Christian soul was capable. As soon as he had grasped these conceptions, which will be found more fully developed in one of the later chapters of his book, the spell of unbelief was broken.

But, once broken, it was dissolved utterly and entirely; he could allow himself to contemplate fearlessly all sorts of issues from which one whose experiences had been less varied would have shrunk. He was free of the enemy's camp, and could go hither and thither whithersoever he would. The very points which to others were insuperable difficulties were to him foundation-stones of faith. For example, to the objection that if in the present state of the records no clear conception of the nature of Christ's life and teaching could be formed, we should be compelled to take one for our model of whom we knew little or nothing certain, I have heard him answer, "And so much the better for us all. The truth, if read by the light of man's imperfect understanding, would have been falser to him than any falsehood. It would have been truth no longer. Better be led aright by an error which is so adjusted as to compensate for the errors in man's powers of understanding, than be misled by a truth which can never be translated from objectivity to subjectivity. In such a case, it is the error which is the truth and the truth the error."

Fearless himself, he could not understand the fears felt by others; and this was perhaps his greatest sympathetic weakness. He was impatient of the subterfuges with which untenable interpretations of Scripture were defended, and of the disingenuousness of certain harmonists; indeed, the mention of the word harmony was enough to kindle an outbreak of righteous anger, which would sometimes go to the utmost limit of righteousness. "Harmonies!" he would exclaim, "the sweetest harmonies are those which are most full of discords, and the discords of one generation of musicians become heavenly music in the hands of their successors. Which of the great musicians has not enriched his art not only by the discovery of new harmonies, but by proving that sounds which are actually inharmonious are nevertheless essentially and eternally delightful? What an outcry has there not always been against the 'unwarrantable licence' with the rules of harmony whenever a Beethoven or a Mozart has broken through any of the trammels which have been regarded as the safeguards of the art, instead of in their true light of fetters, and how gratefully have succeeding musicians acquiesced in and adopted the innovation." Then would follow a tirade with illustration upon illustration, comparison of this passage with that, and an exhaustive demonstration that one or other, or both, could have had no sort of possible foundation in fact; he could only see that the persons from whom he differed were defending something which was untrue and which they ought to have known to be untrue, but he could not see that people ought to know many things which they do not know.

Had he himself seen all that he ought to have been able to see from his own standpoints? Can any of us do so? The force of early bias and education, the force of intellectual surroundings, the force of natural timidity, the force of dulness, were things which he could appreciate and make allowance for in any other age, and among any other people than his own; but as belonging to England and the Nineteenth Century they had no place in his theory of Nature; they were inconceivable, unnatural, unpardonable, whenever they came into contact with the subject of Christian evidences. Deplorable, indeed, they are, but this was just the sort of word to which he could not confine himself. The criticisms upon the late Dean Alford's notes, which will be given in the sequel, display this sort of temper; they are not entirely his own, but he adopted them and endorsed them with a warmth which we cannot but feel to be unnecessary, not to say more. Yet I am free to confess that whatever editorial licence I could venture to take has been taken in the direction of lenity.

On the whole, however, he valued Dean Alford's work very highly, giving him great praise for the candour with which he not unfrequently set the harmonists aside. For example, in his notes upon the discrepancies between St. Luke's and St. Matthew's accounts of the early life of our Lord, the Dean openly avows that it is quite beyond his purpose to attempt to reconcile the two. "This part of the Gospel history," he writes, "is one where the harmonists, by their arbitrary reconcilement of the two accounts, have given great advantage to the enemies of the faith. As the two accounts now stand, it is wholly impossible to suggest any satisfactory method of uniting them, every one who has attempted it has in some part or other of his hypothesis violated probability and common sense," but in spite of this, the Dean had no hesitation in accepting both the accounts. With reference to this the author of The Jesus of History (Williams and Norgate, 1866)—a work to which my brother admitted himself to be under very great obligations, and which he greatly admired, in spite of his utter dissent from the main conclusion arrived at, has the following note:—

"Dean Alford, N.T. for English readers, admits that the narratives as they stand are contradictory, but he believes both. He is even severe upon the harmonists who attempt to frame schemes of reconciliation between the two, on account of the triumph they thus furnish to the 'enemies of the faith,' a phrase which seems to imply all who believe less than he does. The Dean, however, forgets that the faith which can believe two (apparently) contradictory propositions in matters of fact is a very rare gift, and that for one who is so endowed there are thousands who can be satisfied with a plausible though demonstrably false explanation. To the latter class the despised harmonists render a real service."

Upon this note my brother was very severe. In a letter, dated Dec. 18, 1866, addressed to a friend who had alluded to it, and expressed his concurrence with it as in the main just, my brother wrote: "You are wrong about the note in The Jesus of History, there is more of the Christianity of the future in Dean Alford's indifference to the harmony between the discordant accounts of Luke and Matthew than there would have been even in the most convincing and satisfactory explanation of the way in which they came to differ. No such explanation is possible; both the Dean and the author of The Jesus of History were very well aware of this, but the latter is unjust in assuming that his opponent was not alive to the absurdity of appearing to believe two contradictory propositions at one and the same time. The Dean takes very good care that he shall not appear to do this, for it is perfectly plain to any careful reader that he must really believe that one or both narratives are inaccurate, inasmuch as the differences between them are too great to allow of reconciliation by a supposed suppression of detail.

"This, though not said so clearly as it should have been, is yet virtually implied in the admission that no sort of fact which could by any possibility be admitted as reconciling them had ever occurred to human ingenuity; what, then, Dean Alford must have really felt was that the spiritual value of each account was no less precious for not being in strict accordance with the other; that the objective truth lies somewhere between them, and is of very little importance, being long dead and buried, and living in its results only, in comparison with the subjective truth conveyed by both the narratives, which lives in our hearts independently of precise knowledge concerning the actual facts. Moreover, that though both accounts may perhaps be inaccurate, yet that a very little natural inaccuracy on the part of each writer would throw them apparently very wide asunder, that such inaccuracies are easily to be accounted for, and would, in fact, be inevitable in the sixty years of oral communication which elapsed between the birth of our Lord and the writing of the first Gospel, and again in the eighty or ninety years prior to the third, so that the details of the facts connected with the conception, birth, genealogy, and earliest history of our Saviour are irrecoverable—a general impression being alone possible, or indeed desirable.

"It might perhaps have been more satisfactory if Dean Alford had expressed the above more plainly; but if he had done this, who would have read his book? Where would have been that influence in the direction of truly liberal Christianity which has been so potent during the last twenty years? As it was, the freedom with which the Dean wrote was the cause of no inconsiderable scandal. Or, again, he may not have been fully conscious of his own position: few men are; he had taken the right one, but more perhaps by spiritual instinct than by conscious and deliberate exercise of his intellectual faculties. Finally, compromise is not a matter of good policy only, it is a solemn duty in the interests of Christian peace, and this not in minor matters only—we can all do this much—but in those concerning which we feel most strongly, for here the sacrifice is greatest and most acceptable to God. There are, of course, limits to this, and Dean Alford may have carried compromise too far in the present instance, but it is very transparent. The narrowness which leads the author of The Jesus of History to strain at such a gnat is the secret of his inability to accept the divinity and miracles of our Lord, and has marred the most exhaustively critical exegesis of the life and death of our Saviour with an impotent conclusion."

It is strange that one who could write thus should occasionally have shown himself so little able to apply his own principles. He seems to have been alternately under the influence of two conflicting spirits—at one time writing as though there were nothing precious under the sun except logic, consistency, and precision, and breathing fire and smoke against even very trifling deviations from the path of exact criticism—at another, leading the reader almost to believe that he disregarded the value of any objective truth, and speaking of endeavour after accuracy in terms that are positively contemptuous. Whenever he was in the one mood he seemed to forget the possibility of any other; so much so that I have sometimes thought that he did this deliberately and for the same reasons as those which led Adam Smith to exclude one set of premises in his Theory of Moral Sentiments and another in his Wealth of Nations. I believe, however, that the explanation lies in the fact that my brother was inclined to underrate the importance of belief in the objective truth of any other individual features in the life of our Lord than his Resurrection and Ascension. All else seemed dwarfed by the side of these events. His whole soul was so concentrated upon the centre of the circle that he forgot the circumference, or left it out of sight. Nothing less than the strictest objective truth as to the main facts of the Resurrection and Ascension would content him; the other miracles and the life and teaching of our Lord might then be left open; whatever view was taken of them by each individual Christian was probably the one most desirable for the spiritual wellbeing of each.

Even as regards the Resurrection and Ascension, he did not greatly value the detail. Provided these facts were so established that they could never henceforth be controverted, he thought that the less detail the broader and more universally acceptable would be the effect. Hence, when Dean Alford's notes seemed to jeopardise the evidences for these things, he could brook no trifling; for unless Christ actually died and actually came to life again, he saw no escape from an utter denial of any but natural religion. Christ would have been no more to him than Socrates or Shakespeare, except in so far as his teaching was more spiritual. The triune nature of the Deity—the Resurrection from the dead—the hope of Heaven and salutary fear of Hell—all would go but for the Resurrection and Ascension of Jesus Christ; nothing would remain except a sense of the Divine as a substitute for God, and the current feeling of one's peers as the chief moral check upon misconduct. Indeed, we have seen this view openly advocated by a recent writer, and set forth in the very plainest terms. My brother did not live to see it, but if he had, he would have recognised the fulfilment of his own prophecies as to what must be the inevitable sequel of a denial of our Lord's Resurrection.

It will be seen therefore that he was in no danger of being carried away by a "pet theory." Where light and definition were essential, he would sacrifice nothing of either; but he was jealous for his highest light, and felt "that the whole effect of the Christian scheme was indefinitely heightened by keeping all other lights subordinate"—this at least was the illustration which he often used concerning it. But as there were limits to the value of light and "finding"—limits which had been far exceeded, with the result of an unnatural forcing of the lights, and an effect of garishness and unreality—so there were limits to the as yet unrecognised preciousness of "losing" and obscurity; these limits he placed at the objectivity of our Lord's Resurrection and Ascension. Let there be light enough to show these things, and the rest would gain by being in half-tone and shadow.

His facility of illustration was simply marvellous. From his conversation any one would have thought that he was acquainted with all manner of arts and sciences of which he knew little or nothing. It is true, as has been said already, that he had had some practice in the art of painting, and was an enthusiastic admirer of the masterpieces of Raphael, Titian, Guido, Domenichino, and others; but he could never have been called a painter; for music he had considerable feeling; I think he must have known thorough-bass, but it was hard to say what he did or did not know. Of science he was almost entirely ignorant, yet he had assimilated a quantity of stray facts, and whatever he assimilated seemed to agree with him and nourish his mental being. But though his acquaintance with any one art or science must be allowed to have been superficial only, he had an astonishing perception of the relative bearings of facts which seemed at first sight to be quite beyond the range of one another, and of the relations between the sciences generally; it was this which gave him his felicity and fecundity of illustration—a gift which he never abused. He delighted in its use for the purpose of carrying a clear impression of his meaning to the mind of another, but I never remember to have heard him mistake illustration for argument, nor endeavour to mislead an adversary by a fascinating but irrelevant simile. The subtlety of his mind was a more serious source of danger to him, though I do not know that he greatly lost by it in comparison with what he gained; his sense, however, of distinctions was so fine that it would sometimes distract his attention from points of infinitely greater importance in connection with his subject than the particular distinction which he was trying to establish at the moment.

The reader may be glad to know what my brother felt about retaining the unhistoric passages of Scripture. Would he wish to see them sought for and sifted out? Or, again, what would he propose concerning such of the parables as are acknowledged by every liberal Churchman to be immoral, as, for instance, the story of Dives and Lazarus and the Unjust Steward—parables which can never have been spoken by our Lord, at any rate not in their present shape? And here we have a remarkable instance of his moderation and truly English good sense. "Do not touch one word of them," was his often-repeated exclamation. "If not directly inspired by the mouth of God they have been indirectly inspired by the force of events, and the force of events is the power and manifestation of God; they could not have been allowed to come into their present position if they had not been recognised in the counsels of the Almighty as being of indirect service to mankind; there is a subjective truth conveyed even by these parables to the minds of many, that enables them to lay hold of other and objective truths which they could not else have grasped.

"There can be no question that the communistic utterances of the third gospel, as distinguished from St. Matthew's more spiritual and doubtless more historic rendering of the same teaching, have been of inestimable service to Christianity. Christ is not for the whole only, but also for them that are sick, for the ill-instructed and what we are pleased to call 'dangerous' classes, as well as for the more sober thinkers. To how many do the words, 'Blessed be ye poor: for your's is the kingdom of Heaven' (Luke vi., 20), carry a comfort which could never be given by the 'Blessed are the poor in spirit' of Matthew v., 3. In Matthew we find, 'Blessed are the poor in spirit: for their's is the kingdom of Heaven. Blessed are they that mourn: for they shall be comforted. Blessed are the meek: for they shall inherit the earth. Blessed are they which do hunger and thirst after righteousness: for they shall be filled. Blessed are the merciful: for they shall obtain mercy. Blessed are the pure in heart: for they shall see God. Blessed are the peacemakers: for they shall be called the children of God. Blessed are they which are persecuted for righteousness' sake: for their's is the kingdom of heaven. Blessed are ye, when men shall revile you, and persecute you, and shall say all manner of evil against you falsely, for my sake. Rejoice, and be exceeding glad: for great is your reward in heaven: for so persecuted they the prophets which were before you.' In Luke we read, 'Blessed are ye that hunger now: for ye shall be filled. Blessed are ye that weep now: for ye shall laugh. . . . But woe unto you that are rich! for ye have received your consolation. Woe unto you that are full! for ye shall hunger. Woe unto you that laugh now! for ye shall mourn and weep. Woe unto you, when all men shall speak well of you! for so did their fathers to the false prophets,' where even the grammar of the last sentence, independently of the substance, is such as it is impossible to ascribe to our Lord himself.

"The 'upper' classes naturally turn to the version of Matthew, but the 'lower,' no less naturally to that of Luke, nor is it likely that the ideal of Christ would be one-tenth part so dear to them had not this provision for them been made, not by the direct teaching of the Saviour, but by the indirect inspiration of such events as were seen by the Almighty to be necessary for the full development of the highest ideal of which mankind was capable. All that we have in the New Testament is the inspired word, directly or indirectly, of God, the unhistoric no less than the historic; it is for us to take spiritual sustenance from whatever meats we find prepared for us, not to order the removal of this or that dish; the coarser meats are for the coarser natures; as they grow in grace they will turn from these to the finer: let us ourselves partake of that which we find best suited to us, but do not let us grudge to others the provision that God has set before them. There are many things which though not objectively true are nevertheless subjectively true to those who can receive them; and subjective truth is universally felt to be even higher than objective, as may be shown by the acknowledged duty of obeying our consciences (which is the right to us) rather than any dictate of man however much more objectively true. It is that which is true to us that we are bound each one of us to seek and follow."

Having heard him thus far, and being unable to understand, much less to sympathise with teaching so utterly foreign to anything which I had heard elsewhere, I said to him, "Either our Lord did say the words assigned to him by St. Luke or he did not. If he did, as they stand they are bad, and any one who heard them for the first time would say that they were bad; if he did not, then we ought not to allow them to remain in our Bibles to the misleading of people who will thus believe that God is telling them what he never did tell them—to the misleading of the poor, whom even in low self-interest we are bound to instruct as fully and truthfully as we can."

He smiled and answered, "That is the Peter Bell view of the matter. I thought so once, as, indeed, no one can know better than yourself."

The expression upon his face as he said this was sufficient to show the clearness of his present perception, nevertheless I was anxious to get to the root of the matter, and said that if our Lord never uttered these words their being attributed to him must be due to fraud; to pious fraud, but still to fraud.

"Not so," he answered, "it is due to the weakness of man's powers of memory and communication, and perhaps in some measure to unconscious inspiration. Moreover, even though wrong of some sort may have had its share in the origin of certain of the sayings ascribed to our Saviour, yet their removal now that they have been consecrated by time would be a still greater wrong. Would you defend the spoliation of the monasteries, or the confiscation of the abbey lands? I take it no—still less would you restore the monasteries or take back the lands; a consecrated change becomes a new departure; accept it and turn it to the best advantage. These are things to which the theory of the Church concerning lay baptism is strictly applicable. Fieri non debet, factum valet. If in our narrow and unsympathetic strivings after precision we should remove the hallowed imperfections whereby time has set the glory of his seal upon the gospels as well as upon all other aged things, not for twenty generations will they resume that ineffable and inviolable aspect which our fussy meddlesomeness will have disturbed. Let them alone. It is as they stand that they have saved the world.

"No change is good unless it is imperatively called for. Not even the Reformation was good; it is good now; I acquiesce in it, as I do in anything which in itself not vital has received the sanction of many generations of my countrymen. It is sanction which sanctifieth in matters of this kind. I would no more undo the Reformation now than I would have helped it forward in the sixteenth century. Leave the historic, the unhistoric, and the doubtful to grow together until the harvest: that which is not vital will perish and rot unnoticed when it has ceased to have vitality; it is living till it has done this. Note how the very passages which you would condemn have died out of the regard of any but the poor. Who quotes them? Who appeals to them? Who believes in them? Who indeed except the poorest of the poor attaches the smallest weight to them whatever? To us they are dead, and other passages will die to us in like manner, noiselessly and almost imperceptibly, as the services for the fifth of November died out of the Prayer Book. One day the fruit will be hanging upon the tree, as it has hung for months, the next it will be lying upon the ground. It is not ripe until it has fallen of itself, or with the gentlest shaking; use no violence towards it, confident that you cannot hurry the ripening, and that if shaken down unripe the fruit will be worthless. Christianity must have contained the seeds of growth within itself, even to the shedding of many of its present dogmas. If the dogmas fall quietly in their maturity, the precious seed of truth (which will be found in the heart of every dogma that has been able to take living hold upon the world's imagination) will quicken and spring up in its own time: strike at the fruit too soon and the seed will die."

I should be sorry to convey an impression that I am responsible for, or that I entirely agree with, the defence of the unhistoric which I have here recorded. I have given it in my capacity of editor and in some sort biographer, but am far from being prepared to maintain that it is likely, or indeed ought, to meet with the approval of any considerable number of Christians. But, surely, in these days of self-mystification it is refreshing to see the boldness with which my brother thought, and the freedom with which he contemplated all sorts of issues which are too generally avoided. What temptation would have been felt by many to soften down the inconsistencies and contradictions of the Gospels. How few are those who will venture to follow the lead of scientific criticism, and admit what every scholar must well know to be indisputable. Yet if a man will not do this, he shows that he has greater faith in falsehood than in truth.

CHAPTER III

On my brother's death I came into possession of several of his early commonplace books filled with sketches for articles; some of these are more developed than others, but they are all of them fragmentary. I do not think that the reader will fail to be interested with the insight into my brother's spiritual and intellectual progress which a few extracts from these writings will afford, and have therefore, after some hesitation, decided in favour of making them public, though well aware that my brother would never have done so. They are too exaggerated to be dangerous, being so obviously unfair as to carry their own antidote. The reader will not fail to notice the growth not only in thought but also in literary style which is displayed by my brother's later writings.

In reference to the very subject of the parables above alluded to, he had written during his time of unbelief:—"Why are we to interpret so literally all passages about the guilt of unbelief, and insist upon the historical character of every miraculous account, while we are indignant if any one demands an equally literal rendering of the precepts concerning human conduct? He that hath two coats is not to give to him that hath none: this would be 'visionary,' 'utopian,' 'wholly unpractical,' and so forth. Or, again, he that is smitten on the one cheek is not to turn the other to the smiter, but to hand the offender over to the law; nor are the commands relative to indifference as to the morrow and a neglect of ordinary prudence to be taken as they stand; nor yet the warnings against praying in public; nor can the parables, any one of them, be interpreted strictly with advantage to human welfare, except perhaps that of the Good Samaritan; nor the Sermon on the Mount, save in such passages as were already the common property of mankind before the coming of Christ. The parables which every one praises are in reality very bad: the Unjust Steward, the Labourers in the Vineyard, the Prodigal Son, Dives and Lazarus, the Sower and the Seed, the Wise and Foolish Virgins, the Marriage Garment, the Man who planted a Vineyard, are all either grossly immoral, or tend to engender a very low estimate of the character of God—an estimate far below the standard of the best earthly kings; where they are not immoral, or do not tend to degrade the character of God, they are the merest commonplaces imaginable, such as one is astonished to see people accept as having been first taught by Christ. Such maxims as those which inculcate conciliation and a forgiveness of injuries (wherever practicable) are certainly good, but the world does not owe their discovery to Christ, and they have had little place in the practice of his followers.

"It is impossible to say that as a matter of fact the English people forgive their enemies more freely now than the Romans did, we will say in the time of Augustus. The value of generosity and magnanimity was perfectly well known among the ancients, nor do these qualities assume any nobler guise in the teaching of Christ than they did in that of the ancient heathen philosophers. On the contrary, they have no direct equivalent in Christian thought or phraseology. They are heathen words drawn from a heathen language, and instinct with the same heathen ideas of high spirit and good birth as belonged to them in the Latin language; they are no part or parcel of Christianity, and are not only independent of it, but savour distinctly of the flesh as opposed to the spirit, and are hence more or less antagonistic to it, until they have undergone a certain modification and transformation—until, that is to say, they have been mulcted of their more frank and genial elements. The nearest approach to them in Christian phrase is 'self-denial,' but the sound of this word kindles no smile of pleasure like that kindled by the ideas of generosity and nobility of conduct. At the thought of self-denial we feel good, but uncomfortable, and as though on the point of performing some disagreeable duty which we think we ought to pretend to like, but which we do not like. At the thought of generosity, we feel as one who is going to share in a delightfully exhilarating but arduous pastime—full of the most pleasurable excitement. On the mention of the word generosity we feel as if we were going out hunting; at the word 'self-denial,' as if we were getting ready to go to church. Generosity turns well-doing into a pleasure, self-denial into a duty, as of a servant under compulsion.

"There are people who will deny this, but there are people who will deny anything. There are some who will say that St. Paul would not have condemned the Falstaff plays, Twelfth Night, The Tempest, A Midsummer Night's Dream, and almost everything that Shakspeare ever wrote; but there is no arguing against this. 'Every man,' said Dr. Johnson, 'has a right to his own opinion, and every one else has a right to knock him down for it.' But even granting that generosity and high spirit have made some progress since the days of Christ, allowance must be made for the lapse of two thousand years, during which time it is only reasonable to suppose that an advance would have been made in civilisation—and hence in the direction of clemency and forbearance—whether Christianity had been preached or not, but no one can show that the modern English, if superior to the ancients in these respects, show any greater superiority than may be ascribed justly to centuries of established order and good government."

"Again, as to the ideal presented by the character of Christ, about which so much has been written; is it one which would meet with all this admiration if it were presented to us now for the first time? Surely it offers but a peevish view of life and things in comparison with that offered by other highest ideals—the old Roman and Greek ideals, the Italian ideal, and the Shakespearian ideal."

"As with the parables so with the Sermon on the Mount—where it is not commonplace it is immoral, and vice versâ; the admiration which is so freely lavished upon the teachings of Jesus Christ turns out to be but of the same kind as that bestowed upon certain modern writers, who have made great reputations by telling people what they perfectly well knew; and were in no particular danger of forgetting. There is, however, this excuse for those who have been carried away with such musical but untruthful sentences as 'Blessed are they that mourn: for they shall be comforted,' namely, that they have not come to the subject with unbiassed minds. It is one thing to see no merit in a picture, and another to see no merit in a picture when one is told that it is by Raphael; we are few of us able to stand against the prestige of a great name; our self-love is alarmed lest we should be deficient in taste, or, worse still, lest we should be considered to be so; as if it could matter to any right-minded person whether the world considered him to be of good taste or not, in comparison with the keeping of his own soul truthful to itself.

"But if this holds good about things which are purely matters of taste, how much more does it do so concerning those who make a distinct claim upon us for moral approbation or the reverse? Such a claim is most imperatively made by the teaching of Jesus Christ: are we then content to answer in the words of others—words to which we have no title of our own—or shall we strip ourselves of preconceived opinion, and come to the question with minds that are truly candid? Whoever shrinks from this is a liar to his own self, and as such, the worst and most dangerous of liars. He is as one who sits in an impregnable citadel and trembles in a time of peace—so great a coward as not even to feel safe when he is in his own keeping. How loose of soul if he knows that his own keeping is worthless, how aspen-hearted if he fears lest others should find him out and hurt him for communing truthfully with himself!

"That a man should lie to others if he hopes to gain something considerable—this is reckoned cheating, robbing, fraudulent dealing, or whatever it may be; but it is an intelligible offence in comparison with the allowing oneself to be deceived. So in like manner with being bored. The man who lets himself be bored is even more contemptible than the bore. He who puts up with shoddy pictures, shoddy music, shoddy morality, shoddy society, is more despicable than he who is the prime agent in any of these things. He has less to gain, and probably deceives himself more; so that he commits the greater crime for the less reward. And I say emphatically that the morality which most men profess to hold as a Divine revelation was a shoddy morality, which would neither wash nor wear, but was woven together from a tissue of dreams and blunders, and steeped in blood more virulent than the blood of Nessus.

"Oh! if men would but leave off lying to themselves! If they would but learn the sacredness of their own likes and dislikes, and exercise their moral discrimination, making it clear to themselves what it is that they really love and venerate. There is no such enemy to mankind as moral cowardice. A downright vulgar self-interested and unblushing liar is a higher being than the moral cur whose likes and dislikes are at the beck and call of bullies that stand between him and his own soul; such a creature gives up the most sacred of all his rights for something more unsubstantial than a mess of pottage—a mental serf too abject even to know that he is being wronged. Wretched emasculator of his own reason, whose jejune timidity and want of vitality are thus omnipresent in the most secret chambers of his heart!

"We can forgive a man for almost any falsehood provided we feel that he was under strong temptation and well knew that he was deceiving. He has done wrong—still we can understand it, and he may yet have some useful stuff about him—but what can we feel towards one who for a small motive tells lies even to himself, and does not know that he is lying? What useless rotten fig-wood lumber must not such a thing be made of, and what lies will there not come out of it, falling in every direction upon all who come within its reach. The common self-deceiver of modern society is a more dangerous and contemptible object than almost any ordinary felon, a matter upon which those who do not deceive themselves need no enlightenment."

"But why insist so strongly on the literal interpretation of one part of the sayings of Christ, and be so elastic about that of the passages which inculcate more than those ordinary precepts which all had agreed upon as early as the days of Solomon and probably earlier? We have cut down Christianity so as to make it appear to sanction our own conventions; but we have not altered our conventions so as to bring them into harmony with Christianity. We do not give to him that asketh; we take good care to avoid him; yet if the precept meant only that we should be liberal in assisting others—it wanted no enforcing: the probability is that it had been enforced too much rather than too little already; the more literally it has been followed the more terrible has the mischief been; the saying only becomes harmless when regarded as a mere convention. So with most parts of Christ's teaching. It is only conventional Christianity which will stand a man in good stead to live by; true Christianity will never do so. Men have tried it and found it fail; or, rather, its inevitable failure was so obvious that no age or country has ever been mad enough to carry it out in such a manner as would have satisfied its founders. So said Dean Swift in his Argument against abolishing Christianity. 'I hope,' he writes, 'no reader imagines me so weak as to stand up in defence of real Christianity, such as used in primitive times' (if we may believe the authors of those ages) 'to have an influence upon men's beliefs and actions. To offer at the restoring of that would be, indeed, a wild project; it would be to dig up foundations, to destroy at one blow all the wit and half the learning of the kingdom, to break the entire frame and constitution of things, to ruin trade, extinguish arts and sciences, with the professors of them; in short, to turn our courts of exchange and shops into deserts; and would be full as absurd as the proposal of Horace where he advises the Romans all in a body to leave their city, and to seek a new seat in some remote part of the world by way of cure for the corruption of their manners.

"'Therefore, I think this caution was in itself altogether unnecessary (which I have inserted only to prevent all possibility of cavilling), since every candid reader will easily understand my discourse to be intended only in defence of nominal Christianity, the other having been for some time wholly laid aside by general consent as utterly inconsistent with our present schemes of wealth and power.'

"Yet but for these schemes of wealth and power the world would relapse into barbarianism; it is they and not Christianity which have created and preserved civilisation. And what if some unhappy wretch, with a serious turn of mind and no sense of the ridiculous, takes all this talk about Christianity in sober earnest, and tries to act upon it? Into what misery may he not easily fall, and with what life-long errors may he not embitter the lives of his children!

"Again, we do not cut off our right hand nor pluck out our eyes if they offend us; we conventionalise our interpretations of these sayings at our will and pleasure; we do take heed for the morrow, and should be inconceivably wicked and foolish were we not to do so; we do gather up riches, and indeed we do most things which the experience of mankind has taught us to be to our advantage, quite irrespectively of any precept of Christianity for or against. But why say that it is Christianity which is our chief guide, when the words of Christ point in such a very different direction from that which we have seen fit to take? Perhaps it is in order to compensate for our laxity of interpretation upon these points that we are so rigid in stickling for accuracy upon those which make no demand upon our comfort or convenience? Thus, though we conventionalise practice, we never conventionalise dogma. Here, indeed, we stickle for the letter most inflexibly; yet one would have thought that we might have had greater licence to modify the latter than the former. If we say that the teaching of Christ is not to be taken according to its import—why give it so much importance? Teaching by exaggeration is not a satisfactory method, nor one worthy of a being higher than man; it might have been well once, and in the East, but it is not well now. It induces more and more of that jarring and straining of our moral faculties, of which much is unavoidable in the existing complex condition of affairs, but of which the less the better. At present the tug of professed principles in one direction, and of necessary practice in the other, causes the same sort of wear and tear in our moral gear as is caused to a steam-engine by continually reversing it when it is going it at full speed. No mechanism can stand it."

The above extracts (written when he was about twenty-three years old) may serve to show how utter was the subversion of his faith. His mind was indeed in darkness! Who could have hoped that so brilliant a day should have succeeded to the gloom of such mistrust? Yet as upon a winter's morning in November when the sun rises red through the smoke, and presently the fog spreads its curtain of thick darkness over the city, and then there comes a single breath of wind from some more generous quarter, whereupon the blessed sun shines again, and the gloom is gone; or, again, as when the warm south-west wind comes up breathing kindness from the sea, unheralded, suspected, when the earth is in her saddest frost, and on the instant all the lands are thawed and opened to the genial influences of a sweet springful whisper—so thawed his heart, and the seed which had lain dormant in its fertile soil sprang up, grew, ripened, and brought forth an abundant harvest.

Indeed now that the result has been made plain we can perhaps feel that his scepticism was precisely of that nature which should have given the greatest ground for hope. He was a genuine lover of truth in so far as he could see it.

His lights were dim, but such as they were he walked according to them, and hence they burnt ever more and more clearly, till in later life they served to show him what is vouchsafed to such men and to such only—the enormity of his own mistakes. Better that a man should feel the divergence between Christian theory and Christian practice, that he should be shocked at it—even to the breaking away utterly from the theory until he has arrived at a wider comprehension of its scope—than that he should be indifferent to the divergence and make no effort to bring his principles and practice into harmony with one another. A true lover of consistency, it was intolerable to him to say one thing with his lips and another with his actions. As long as this is true concerning any man, his friends may feel sure that the hand of the Lord is with him, though the signs thereof be hidden from mortal eyesight.

CHAPTER IV

During the dark and unhappy time when he had, as it seems to me, bullied himself, or been bullied into infidelity, he had been utterly unable to realise the importance even of such a self-evident fact as that our Lord addressing an Eastern people would speak in such a way as Eastern people would best understand; it took him years to appreciate this. He could not see that modes of thought are as much part of a language as the grammar and words which compose it, and that before a passage can be said to be translated from one language into another it is often not the words only which must be rendered, but the thought itself which must be transformed; to a people habituated to exaggeration a saying which was not exaggerated would have been pointless—so weak as to arrest the attention of no one; in order to translate it into such words as should carry precisely the same meaning to colder and more temperate minds, the words would often have to be left out of sight altogether, and a new sentence or perhaps even simile or metaphor substituted; this is plainly out of the question, and therefore the best course is that which has been taken, i.e., to render the words as accurately as possible, and leave the reader to modify the meaning. But it was years before my brother could be got to feel this, nor did he ever do so fully, simple and obvious though it must appear to most people, until he had learned to recognise the value of a certain amount of inaccuracy and inconsistency in everything which is not comprehended in mechanics or the exact sciences. "It is this," he used to say, "which gives artistic or spiritual value as contrasted with mechanical precision."

In inaccuracy and inconsistency, therefore (within certain limits), my brother saw the means whereby our minds are kept from regarding things as rigidly and immutably fixed which are not yet fully understood, and perhaps may never be so while we are in our present state of probation. Life is not one of the exact sciences, living is essentially an art and not a science. Every thing addressed to human minds at all must be more or less of a compromise; thus, to take a very old illustration, even the definitions of a point and a line—the fundamental things in the most exact of the sciences—are mere compromises. A point is supposed to have neither length, breadth, nor thickness—this in theory, but in practice unless a point have a little of all these things there is nothing there. So with a line; a line is supposed to have length, but no breadth, yet in practice we never saw a line which had not breadth. What inconsistency is there here, in requiring us to conceive something which we cannot conceive, and which can have no existence, before we go on to the investigation of the laws whereby the earth can alone be measured and the orbits of the planets determined. I do not think that this illustration was presented to my brother's mind while he was young, but I am sure that if it had been it would have made him miserable. He would have had no confidence in mathematics, and would very likely have made a furious attack upon Newton and Galileo, and been firmly convinced that he was discomfiting them. Indeed I cannot forget a certain look of bewilderment which came over his face when the idea was put before him, I imagine, for the first time. Fortunately he had so grown that the right inference was now in no danger of being missed. He did not conclude that because the evidences for mathematics were founded upon compromises and definitions which are inaccurate—therefore that mathematics were false, or that there were no mathematics, but he learnt to feel that there might be other things which were no less indisputable than mathematics, and which might also be founded on facts for which the evidences were not wholly free from inconsistencies and inaccuracies.

To some he might appear to be approaching too nearly to the "Sed tu vera puta" argument of Juvenal. I greatly fear that an attempt may be made to misrepresent him as taking this line; that is to say, as accepting Christianity on the ground of the excellence of its moral teaching, and looking upon it as, indeed, a superstition, but salutary for women and young people. Hardly anything would have shocked him more profoundly. This doctrine with its plausible show of morality appeared to him to be, perhaps, the most gross of all immoralities, inasmuch as it cuts the ground from under the feet of truth, luring the world farther and farther from the only true salvation—the careful study of facts and of the safest inferences that may be drawn from them. Every fact was to him a part of nature, a thing sacred, pregnant with Divine teaching of some sort, as being the expression of Divine will. It was through facts that he saw God; to tamper with facts was, in his view, to deface the countenance of the Almighty. To say that such and such was so and so, when the speaker did not believe it, was to lead people to worship a false God instead of a true one; an ειδωλον; setting them, to quote the words of the Psalmist, "a-whoring after their own imaginations." He saw the Divine presence in everything—the evil as well as the good; the evil being the expression of the Divine will that such and such courses should not go unpunished, but bring pain and misery which should deter others from following them, and the good being his sign of approbation. There was nothing good for man to know which could not be deduced from facts. This was the only sound basis of knowledge, and to found things upon fiction which could be made to stand upon facts was to try and build upon a quicksand.

He, therefore, loathed the reasoning of Juvenal with all the intensity of his nature. It was because he believed that the Resurrection and Ascension of our Lord were just as much matters of actual history as the assassination of Julius Cæsar, and that they happened precisely in the same way as every daily event happens at present—that he accepted the Christian scheme in its essentials. Then came the details. Were these also objectively true? He answered, "Certainly not in every case." He would not for the world have had any one believe that he so considered them; but having made it perfectly clear that he was not going to deceive himself, he set himself to derive whatever spiritual comfort he could from them, just as he would from any noble fiction or work of art, which, while not professing to be historical, was instinct with the soul of genius. That there were unhistorical passages in the New Testament was to him a fact; therefore it was to be studied as an expression of the Divine will. What could be the meaning of it? That we should consider them as true? Assuredly not this. Then what else? This—that we should accept as subjectively true whatever we found spiritually precious, and be at liberty to leave all the rest alone—the unhistoric element having been introduced purposely for the sake of giving greater scope and latitude to the value of the ideal.

Of course one who was so firmly persuaded of the objective truth of the Resurrection and Ascension could be in no sort of danger of relapsing into infidelity as long as his reason remained. During the years of his illness his mind was clearly impaired, and no longer under his own control; but while his senses were his own it was absolutely impossible that he could be shaken by discrepancies and inconsistencies in the gospels. What small and trifling things are such discrepancies by the side of the great central miracle of the Resurrection! Nevertheless their existence was indisputable, and was no less indisputably a cause of stumbling to many, as it had been to himself. His experience of his own sufferings as an unbeliever gave him a keener sympathy with those who were in that distressing condition than could be felt by any one who had not so suffered, and fitted him, perhaps, more than any one who has yet lived to be the interpreter of Christianity to the Rationalist, and of Rationalism to the Christian. This, accordingly, was the task to which he set himself, having been singularly adapted for it by Nature, and as singularly disciplined by events.

It seemed to him that the first thing was to make the two parties understand one another—a thing which had never yet been done, but which was not at all impossible. For Protestantism is raised essentially upon a Rationalistic base. When we come to a definition of Rationalism nothing can be plainer than that it demands no scepticism from any one which an English Protestant would not approve of. It is another matter with the Church of Rome. That Church openly declares it as an axiom that religion and reason have nothing to do with one another, and that religion, though in flat contradiction to reason, should yet be accepted from the hands of a certain order as an act of unquestioning faith. The line of separation therefore between the Romanist and the Rationalist is clear, and definitely bars any possibility of arrangement between the two. Not so with the Protestant, who as heartily as the Rationalist admits that nothing is required to be believed by man except such things as can be reasonably proved—i.e., proved to the satisfaction of the reason. No Protestant would say that the Christian scheme ought to be accepted in spite of its being contrary to reason; we say that Christianity is to be believed because it can be shewn to follow as the necessary consequence of using our reason rightly. We should be shocked at being supposed to maintain otherwise. Yet this is pure Rationalism. The Rationalist would require nothing more; he demurs to Christianity because he maintains that if we bring our reason to bear upon the evidences which are brought forward in support of it, we are compelled to reject it; but he would accept it without hesitation if he believed that it could be sustained by arguments which ought to carry conviction to the reason. Thus both are agreed in principle that if the evidences of Christianity satisfy human reason, then Christianity should be received, but that on any other supposition it should be rejected.

Here then, he said, we have a common starting-point and the main principle of Rationalism turns out to be nothing but what we all readily admit, and with which we and our fathers have been as familiar for centuries as with the air we breathe. Every Protestant is a Rationalist, or else he ought to be ashamed of himself. Does he want to be called an "Irrationalist"? Hardly—yet if he is not a Rationalist what else can he be? No: the difference between us is one of detail, not of principle. This is a great step gained.

The next thing therefore was to make each party understand the view which the other took concerning the position which they had agreed to hold in common. There was no work, so far as he knew, which would be accepted both by Christians and unbelievers as containing a fair statement of the arguments of the two contending parties: every book which he had yet seen upon either side seemed written with the view of maintaining that its own side could hold no wrong, and the other no right: neither party seemed to think that they had anything to learn from the other, and neither that any considerable addition to their knowledge of the truth was either possible or desirable. Each was in possession of truth already, and all who did not see and feel this must be either wilfully blinded, or intensely stupid, or hypocrites.

So long as people carried on a discussion thus, what agreement was possible between them? Yet where, upon the Christian side, was the attempt to grapple with the real difficulties now felt by unbelievers? Simply nowhere. All that had been done hitherto was antiquated. Modern Christianity seemed to shrink from grappling with modern Rationalism, and displayed a timidity which could not be accounted for except by the supposition of secret misgiving that certain things were being defended which could not be defended fairly. This was quite intolerable; a misgiving was a warning voice from God, which should be attended to as a man valued his soul. On the other hand, the conviction reasonably entertained by unbelievers that they were right on many not inconsiderable details of the dispute, and that so-called orthodox Christians in their hearts knew it but would not own it—or that if they did not know it, they were only in ignorance because it suited their purpose to be so—this conviction gave an overweening self-confidence to infidels, as though they must be right in the whole because they were so in part; they therefore blinded themselves to all the more fundamental arguments in support of Christianity, because certain shallow ones had been put forward in the front rank, and been far too obstinately defended. They thus regarded the question too superficially, and had erred even more through pride of intellect and conceit than their opponents through timidity.

What then was to be done? Surely this; to explain the two contending parties to one another; to show to Rationalists that Christians are right upon Rationalistic principles in all the more important of their allegations; that is to say, to establish the Resurrection and Ascension of the Redeemer upon a basis which should satisfy the most imperious demands of modern criticism. This would form the first and most important part of the task. Then should follow a no less convincing proof that Rationalists are right in demurring to the historical accuracy of much which has been too obstinately defended by so-called orthodox writers. This would be the second part. Was there not reason to hope that when this was done the two parties might understand one another, and meet in a common Christianity? He believed that there was, and that the ground had been already cleared for such mutual compromise as might be accepted by both sides, not from policy but conviction. Therefore he began writing the book which it has devolved upon myself to edit, and which must now speak for itself. For him it was to suffer and to labour; almost on the very instant of his having done enough to express his meaning he was removed from all further power of usefulness.

The happy change from unbelief to faith had already taken place some three or four years before my return from America. With it had also come that sudden development of intellectual and spiritual power which so greatly astonished even those who had known him best. The whole man seemed changed—to have become possessed of an unusually capacious mind, instead of one which was acute, but acute only. On looking over the earlier letters which I received from him when I was in America, I can hardly believe that they should have been written by the same person as the one to whom, in spite of not a few great mental defects, I afterwards owed more spiritual enrichment than I have owed to any other person. Yet so it was. It came upon me imperceptibly that I had been very stupid in not discovering that my brother was a genius; but hardly had I made the discovery, and hardly had the fragment which follows this memoir received its present shape, when his overworked brain gave way and he fell into a state little better than idiocy. His originally cheerful spirits left him, and were succeeded by a religious melancholy which nothing could disturb. He became incapable either of mental or physical exertion, and was pronounced by the best physicians to be suffering from some obscure disease of the brain brought on by excitement and undue mental tension: in this state he continued for about four years, and died peacefully, but still as one in the profoundest melancholy, on the 15th of March, 1872, aged 40.

Always hopeful that his health would one day be restored, I never ventured to propose that I should edit his book during his own life-time. On his death I found his papers in the most deplorable confusion. The following chapters had alone received anything like a presentable shape—and these providentially are the most essential.

A dream is a dream only, yet sometimes there follows a fulfilment which bears a strange resemblance to the thing dreamt of. No one now believes that the Book of Revelation is to be taken as foretelling events which will happen in the same way as the massacre, for instance, of St. Bartholomew, indeed it is doubtful how far the whole is not to be interpreted as an allegory, descriptive of spiritual revolutions; yet surely my mother's dream as to the future of one, at least, of her sons has been strangely verified, and it is believed that the reader when he lays down this volume will feel that there have been few more potent witnesses to the truth of Christ than John Pickard Owen.

CHAPTER I

INTRODUCTION

It is to be feared that there is no work upon the evidences of our faith, which is as satisfactory in its completeness and convincing power as we have a right to expect when we consider the paramount importance of the subject and the activity of our enemies. Otherwise why should there be no sign of yielding on the part of so many sincere and eminent men who have heard all that has been said upon the Christian side and are yet not convinced by it? We cannot think that the many philosophers who make no secret of their opposition to the Christian religion are unacquainted with the works of Butler and Paley—of Mansel and Liddon. This cannot be: they must be acquainted with them, and find them fail.

Now, granting readily that in some minds there is a certain wilful and prejudiced self-blindness which no reasoning can overcome, and granting also that men very much preoccupied with any one pursuit (more especially a scientific one) will be apt to give but scant and divided attention to arguments upon other subjects such as religion or politics, nevertheless we have so many opponents who profess to have made a serious study of Christian evidences, and against whose opinion no exception can be fairly taken, that it seems as though we were bound either to admit that our demonstrations require rearrangement and reconsideration, or to take the Roman position, and maintain that revelation is no fit subject for evidence but is to be accepted upon authority. This last position will be rejected at once by nine-tenths of Englishmen. But upon rejecting it we look in vain for a work which shall appear to have any such success in arresting infidelity as attended the works of Butler and Paley in the last century. In their own day these two great men stemmed the current of infidelity: but no modern writers have succeeded in doing so, and it will scarcely be said that either Butler or Paley set at rest the many serious and inevitable questions in connection with Christianity which have arisen during the last fifty years. We could hardly expect one of the more intelligent students at Oxford or Cambridge to find his mind set once and for ever free from all rising doubt either by the Analogy or the Evidences. Suppose, for example, that he has been misled by the German writers of the Tübingen school, how will either of the above-named writers help him? On the contrary, they will do him harm, for they will not meet the requirements of the case, and the inference is too readily drawn that nothing else can do so. It need hardly be insisted upon that this inference is a most unfair one, but surely the blame of its being drawn rests in some measure at the door of those whose want of thoroughness has left people under the impression that no more can be said than what has been said already.

It is the object, therefore, of this book to contribute towards establishing Christian evidences upon a more secure and self-evident base than any upon which they are made to rest at present, so far, that is to say, as a work which deliberately excludes whole fields of Christian evidence can tend towards so great a consummation. In spite of the narrow limits within which I have resolved to keep my treatment of the subject, I trust that I may be able to produce such an effect upon the minds of those who are in doubt concerning the evidences for the hope that is in them, that henceforward they shall never doubt again. I am not sanguine enough to suppose that I shall be able to induce certain eminent naturalists and philosophers to reopen a question which they have probably long laid aside as settled; unfortunately it is not in any but the very noblest Christian natures to do this, nevertheless, could they be persuaded to read these pages I believe that they would find so much which would be new to them, that their prejudices would be greatly shaken. To the younger band of scientific investigators I appeal more hopefully.

It may be asked why not have undertaken the whole subject and devoted a life-time to writing an exhaustive work? The answer suggests itself that the believer is in no want of such a book, while the unbeliever would be repelled by its size. Assuredly there can be no doubt as to the value of a great work which should meet objections derived from certain recent scientific theories, and confute opponents who have arisen since the death of our two great apologists, but as a preliminary to this a smaller and more elementary book seems called for, which shall give the main outlines of our position with such boldness and effectiveness as to arrest the attention of any unbeliever into whose hands it may fall, and induce him to look further into what else may be urged upon the Christian side. We are bound to adapt our means to our ends, and shall have a better chance of gaining the ear of our adversaries if we can offer them a short and pregnant book than if we come to them with a long one from which whole chapters might be pruned. We have to bring the Christian religion to men who will look at no book which cannot be read in a railway train or in an arm-chair; it is most deplorable that this should be the case, nevertheless it is indisputably a fact, and as such must be attended to by all who hope to be of use in bringing about a better state of things. And let me add that never yet was there a time when it so much behoved all who are impressed with the vital power of religion to bestir themselves; for the symptoms of a general indifference, not to say hostility, must be admitted to be widely diffused, in spite of an imposing array of facts which can be brought forward to the contrary; and not only this, but the stream of infidelity seems making more havoc yearly, as it might naturally be expected to do, when met by no new works of any real strength or permanence.

Bearing in mind, therefore, the necessity for prompt action, it seemed best to take the most overwhelming of all miracles—the Resurrection of our Lord Jesus Christ, and show that it can be so substantiated that no reasonable man should doubt it. This I have therefore attempted, and I humbly trust that the reader will feel that I have not only attempted it, but done it, once and for all so clearly and satisfactorily and with such an unflinching examination of the most advanced arguments of unbelievers, that the question can never be raised hereafter by any candid mind, or at any rate not until science has been made to rest on different grounds from those on which she rests at present.

But the truth of our Lord's resurrection having been once established, what need to encumber this book with further evidences of the miraculous element in his ministry? The other miracles can be no insuperable difficulty to one who accepts the Resurrection. It is true that as Christians we cannot dwell too minutely upon every act and incident in the life of the Redeemer, but unhappily we have to deal with those who are not Christians, and must consider rather what we can get them to take than what we should like to give them: "Be ye wise as serpents and harmless as doves," saith the Saviour. A single miracle is as good as twenty, provided that it be well established, and can be shewn to be so: it is here that even the ablest of our apologists have too often failed; they have professed to substantiate the historical accuracy of all the recorded miracles and sayings of our Lord, with a result which is in some instances feeble and conventional, and occasionally even unfair (oh! what suicidal folly is there in even the remotest semblance of unfairness), instead of devoting themselves to throwing a flood of brilliancy upon the most important features and leaving the others to shine out in the light reflected from these. Even granting that some of the miracles recorded of our Lord are apocryphal, what of that? We do not rest upon them: we have enough and more than enough without them, and can afford to take the line of saying to the unbeliever, "Disbelieve this miracle or that if you find that you cannot accept it, but believe in the Resurrection, of which we will put forward such ample proofs that no healthy reason can withstand them, and, having accepted the Resurrection, admit it as the manifestation of supernatural power, the existence of which can thus no longer be denied."

Does not the reader feel that there is a ring of truth and candour about this which must carry more weight with an opponent than any strained defence of such a doubtful miracle as the healing of the impotent man at the pool of Bethesda? We weight ourselves as against our opponents by trying to defend too much; no matter how sound and able the defence of one part of the Christian scheme may have been, its effect is often marred by contiguity with argument which the writer himself must have suspected, or even known, to be ingenious rather than sound: the moment that this is felt in any book its value with an opponent is at an end, for he must be continually in doubt whether the spirit which he has detected here or there may not be existing and at work in a hundred other places where he has not detected it. What carries weight with an antagonist is the feeling that his position has been mastered and his difficulties grasped with thoroughness and candour.

On this point I am qualified to speak from long and bitter experience. I say that want of candour and the failure to grasp the position occupied, however untenably, by unbelievers is the chief cause of the continuance of unbelief. When this cause has been removed unbelief will die a natural death. For years I was myself a believer in nothing beyond the personality and providence of God: yet I feel (not without a certain sense of bitterness, which I know that I should not feel but cannot utterly subdue) that if my first doubts had been met with patient endeavour to understand their nature and if I had felt that the one in whom I confided had been ready to go to the root of the matter, and even to yield up the convictions of a life-time could it be shewn that they were unsafely founded, my doubts would have been resolved in an hour or two's quiet conversation, and would at once have had the effect, which they have only had after long suffering and unrest, of confirming me in my allegiance to Christ. But I was met with anger and impatience. There was an instinct which told me that my opponent had never heard a syllable against his own convictions, and was determined not to hear one: on this I assumed rashly that he must have good reason for his resolution; and doubt ripened into unbelief. Oh! what years of heart-burning and utter drifting followed. Yet when I was at last brought within the influence of one who not only believed all that my first opponent did, but who also knew that the more light was thrown upon it the more clearly would its truth be made apparent—a man who talked with me as though he was anxious that I should convince him if he were in error, not as though bent on making me believe whatever habit and circumstances had imposed as a formula upon himself—my heart softened at once, and the dry places of my soul were watered.

The above may seem too purely personal to warrant its introduction here, yet the experience is one which should not be without its value to others. Its effect upon myself has been to give me an unutterable longing to save others from sufferings like my own; I know so well where it is that, to use a homely metaphor, the shoe pinches. And it is chiefly here—in the fact that the unbeliever does not feel as though we really wanted to understand him. This feeling is in many cases lamentably well founded. No one likes hearing doubt thrown upon anything which he regards as settled beyond dispute, and this, happily, is what most men feel concerning Christianity. Again, indolence or impotence of mind indisposes many to intellectual effort; others are pained by coming into contact with anything which derogates from the glory due to the great sacrifice of Christ, or to his Divine nature, and lastly not a few are withheld by moral cowardice from daring to bestow the pains upon the unbeliever which his condition requires. But from whichever of these sources the disinclination to understand him comes, its effect is equally disastrous to the unbeliever. People do not mind a difference of opinion, if they feel that the one who differs from them has got a firm grasp of their position; or again, if they feel that he is trying to understand them but fails from some defect either of intellect or education, even in this case they are not pained by opposition. What injures their moral nature and hardens their hearts is the conviction that another could understand them if he chose, but does not choose, and yet none the less condemns them. On this they become imbued with that bitterness against Christianity which is noticeable in so many free-thinkers.

Can we greatly wonder? For, sad though the admission be, it is only justice to admit that we Christians have been too often contented to accept our faith without knowing its grounds, in which case it is more by luck than by cunning that we are Christians at all, and our faith will be in continual danger. The greater number even of those who have undertaken to defend the Christian faith have been sadly inclined to avoid a difficulty rather than to face it, unless it is so easy as to be no real difficulty at all. I do not say that this is unnatural, for the Christian writer must be deeply impressed with the sinfulness of unbelief, and will therefore be anxious to avoid raising doubts which will probably never yet have occurred to his reader, and might possibly never do so; nor does there at first sight appear to be much advantage in raising difficulties for the sole purpose of removing them; nevertheless I cannot think that if either Butler or Paley could have foreseen the continuance of unbelief, and the ruin of so many souls whom Christ died to save, they would have been contented to act so almost entirely upon the defensive.

Yet it is impossible not to feel that we in their place should have done as they did. Infidelity was still in its infancy: the nature of the disease was hardly yet understood; and there seemed reason to fear lest it might be aggravated by the very means taken to cure it; it seemed safer therefore in the first instance to confine attention to the matter actually in debate, and leave it to time to suggest a more active treatment should the course first tried prove unsatisfactory. Who can be surprised that the earlier apologists should have felt thus in the presence of an enemy whose novelty made him appear more portentous than he can ever seem to ourselves? They were bound to venture nothing rashly; what they did they did, for their own age, thoroughly; we owe it to their cautious pioneering that we so know the weakness of our opponents and our own strength as to be able to do fearlessly what may well have seemed perilous to our forefathers: nevertheless it is easy to be wise after the event, and to regret that a bolder course was not taken at the outset. If Butler and Paley had fought as men eager for the fray, as men who smelt the battle from afar, it is impossible to believe that infidelity could have lasted as long as it has. What can be done now could have been done just as effectively then, and though we cannot be surprised at the caution shewn at first, we are bound to deplore it as short-sighted.

The question, however, for ourselves is not what dead men might have done better long ago, but what living men and women can do most wisely now; and in answer to it I would say that there is no policy so unwise as fear in a good cause: the bold course is also the wise one; it consists in being on the lookout for objections, in finding the very best that can be found and stating them in their most intelligible form, in shewing what are the logical consequences of unbelief, and thus carrying the war into the enemy's country; in fighting with the most chivalrous generosity and a determination to take no advantage which is not according to the rules of war most strictly interpreted against ourselves, but within such an interpretation showing no quarter. This is the bold course and the true course: it will beget a confidence which can never be felt in the wariness, however well-intentioned, of the old defenders.

Let me, therefore, beg the reader to follow me patiently while I do my best to put before him the main difficulties felt by unbelievers. When he is once acquainted with these he will run in no danger of confirming doubt through his fear in turning away from it in the first instance. How many die hardened unbelievers through the treatment which they have received from those to whom their Christianity has been a matter of circumstances and habit only? Hell is no fiction. Who, without bitter sorrow, can reflect upon the agonies even of a single soul as being due to the selfishness or cowardice of others? Awful thought! Yet it is one which is daily realised in the case of thousands.

In the commonest justice to brethren, however sinful, each one of us who tries to lead them to the Saviour is bound not only to shew them the whole strength of our own arguments, but to make them see that we understand the whole strength of theirs; for men will not seriously listen to those whom they believe to know one side of a question only. It is this which makes the educated infidel so hard to deal with; he knows very well that an intelligent apprehension of the position held by an opponent is indispensable for profitable discussion; but he very rarely meets with this in the case of those Christians who try to argue with him; he therefore soon acquires a habit of avoiding the subject of religion, and can seldom be induced to enter upon an argument which he is convinced can lead to nothing.

He who would cure a disease must first know what it is, and he who would convert an infidel must know what it is that he is to be converted from, as well as what he is to be led to; nothing can be laid hold of unless its whereabouts is known. It is deplorable that such commonplaces should be wanted; but, alas! it is impossible to do without them. People have taken a panic on the subject of infidelity as though it were so infectious that the very nurses and doctors should run away from those afflicted with it; but such conduct is no less absurd than cruel and disgraceful. Infidelity is only infectious when it is not understood. The smallest reflection should suffice to remind us that a faith which has satisfied the most brilliant and profound of human intellects for nearly two thousand years must have had very sure foundations, and that any digging about them for the purpose of demonstrating their depth and solidity, will result, not in their disturbance, but in its being made clear to every eye that they are laid upon a rock which nothing can shake—that they do indeed satisfy every demand of human reason, which suffers violence not from those who accept the scheme of the Christian redemption, but from those who reject it.

This being the case, and that it is so will, I believe, appear with great clearness in the following pages, what need to shrink from the just and charitable course of understanding the nature of what is urged by those who differ from us? How can we hope to bring them to be of one mind in Christ Jesus with ourselves, unless we can resolve their difficulties and explain them? And how can we resolve their difficulties until we know what they are? Infidelity is as a reeking fever den, which none can enter safely without due precautions, but the taking these precautions is within our own power; we can all rely upon the blessed promises of the Saviour that he will not desert us in our hour of need if we will only truly seek him; there is more infidelity in this shrinking and fear of investigation than in almost any open denial of Christ; the one who refuses to examine the doubts felt by another, and is prevented from making any effort to remove them through fear lest he should come to share them, shews either that he has no faith in the power of Christianity to stand examination, or that he has no faith in the promises of God to guide him into all truth. In either case he is hardly less an unbeliever than those whom he condemns.

Let the reader therefore understand that he will here find no attempt to conceal the full strength of the arguments relied on by unbelievers. This manner of substantiating the truth of Christianity has unhappily been tried already; it has been tried and has failed as it was bound to fail. Infidelity lives upon concealment. Shew it in broad daylight, hold it up before the world and make its hideousness manifest to all—then, and not till then, will the hours of unbelief be numbered. We have been the mainstay of unbelief through our timidity. Far be it from me, therefore, that I should help any unbeliever by concealing his case for him. This were the most cruel kindness. On the contrary, I shall insist upon all his arguments and state them, if I may say so without presumption, more clearly than they have ever been stated within the same limits. No one knows what they are better than I do. No one was at one time more firmly persuaded that they were sound. May it be found that no one has so well known how also to refute them.

The reader must not therefore expect to find fictitious difficulties in the way of accepting Christianity set up with one hand in order to be knocked down again with the other: he will find the most powerful arguments against all that he holds most sacred insisted on with the same clearness as those on his own side; it is only by placing the two contending opinions side by side in their utmost development that the strength of our own can be made apparent. Those who wish to cry peace, peace, when there is no peace, those who would take their faith by fashion as the take their clothes, those who doubt the strength of their own cause and do not in their heart of heart believe that Christianity will stand investigation, those, again, who care not who may go to Hell provided they are comfortably sure of going to Heaven themselves, such persons may complain of the line which I am about to take. They on the other hand whose faith is such that it knows no fear of criticism, and they whose love for Christ leads them to regard the bringing of lost souls into his flock as the highest earthly happiness—such will admit gladly that I have been right in tearing aside the veil from infidelity and displaying it uncloaked by the side of faith itself.

At the same time I am bound to confess that I never should have been able to see the expediency, not to say the absolute necessity for such a course, unless I had been myself for many years an unbeliever. It is this experience, so bitterly painful, that has made me feel so strongly as to the only manner in which others can be brought from darkness into light. The wisdom of the Almighty recognised that if man was to be saved it must be done by the assumption of man's nature on the part of the Deity. God must make himself man, or man could never learn the nature and attributes of God. Let us then follow the sublime example of the incarnation, and make ourselves as unbelievers that we may teach unbelievers to believe. If Paley and Butler had only been real infidels for a single year, instead of taking the thoughts and reasonings of their opponents at second-hand, what a difference should we not have seen in the nature of their work. Alas! their clear and powerful intellects had been trained early in the severest exercises; they could not be misled by any of the sophistries of their opponents; but, on the other hand, never having been misled they knew not the thread of the labyrinth as one who has been shut up therein.

I should also warn the reader of another matter. He must not expect to find that I can maintain everything which he could perhaps desire to see maintained. I can prove, to such a high degree of presumption as shall amount virtually to demonstration, that our Lord died upon the cross, rose again from the dead upon the third day, and ascended into Heaven: but I cannot prove that none of the accounts of these events which have come down to us have suffered from the hand of time: on the contrary, I must own that the reasons which led me to conclude that there must be confusion in some of the accounts of the Resurrection continue in full force with me even now. I see no way of escaping from this conclusion: but it seems equally strange that the Christian should have such an indomitable repugnance to accept it, and that the unbeliever should conceive that it inflicts any damage whatever upon the Christian evidences. Perhaps the error of each confirms that of the other, as will appear hereafter.

I have spoken hitherto as though I were writing only for men, but the help of good women can never be so precious as in the salvation of human souls; if there is one work for which women are better fitted than another, it is that of arresting the progress of unbelief. Can there be a nobler one? Their superior tact and quickness give them a great advantage over men; men will listen to them when they would turn away from one of their own sex; and though I am well aware that courtesy is no argument, yet the natural politeness shewn by a man to a woman will compel attention to what falls from her lips, and will thus perhaps be the means of bringing him into contact with Divine truths which would never otherwise have reached him. Yet this is a work from which too many women recoil in horror—they know that they can do nothing unless they are intimately acquainted with the opinions of those from whom they differ, and from such an intimacy they believe that they are right in shrinking.

Oh, my sisters, my sisters, ye who go into the foulest dens of disease and vice, fearless of the pestilence and of man's brutality, ye whose whole lives bear witness to the cross of Christ and the efficacy of the Divine love, did one of you ever fear being corrupted by the vice with which you came in contact? Is there one of you who fears to examine why it is that even the most specious form of vice is vicious? You fear not infection here, for you know that you are on sure ground, and that there is no form of vice of which the viciousness is not clearly provable; but can you doubt that the foundation of your faith is sure also, and can you not see that your cowardice in not daring to examine the foul and soul-destroying den of infidelity is a stumbling-block to those who have not yet known their Saviour? Your fear is as the fear of children who dare not go in the dark; but alas! the unbeliever does not understand it thus. He says that your fear is not of the darkness but of the light, and that you dare not search lest you should find that which would make against you. Hideous blasphemy against the Lord! But is not the sin to be laid partly at the door of those whose cowardice has given occasion for it?

Is there none of you who knows that as to the pure all things are pure, so to the true and loyal heart all things will confirm its faith? You shrink from this last trial of your allegiance, partly from the pain of even seeing the wounds of your Redeemer laid open—of even hearing the words of those enemies who have traduced him and crucified him afresh—but you lose the last and highest of the prizes, for great as is your faith now, be very sure that from this crowning proof of your devotion you would emerge with greater still.

Has none of you seen a savage dog barking and tearing at the end of his chain as though he were longing to devour you, and yet if you have gone bravely up to him and bade him be still, he is cowed and never barks again? Such is the genius of infidelity; it loves to threaten those who retreat, yet it shrinks daunted back from those who meet it boldly; it is the lack of boldness on the part of the Christian which gives it all its power; when Christians are strong in the strength of their own cause infidels will know their impotence, but as long as there are cowards there will be those who prey upon cowardice, and as long as those who should defend the cross of Christ hide themselves behind battlements, so long will the enemy come up to the very walls of the defence and trouble them that are within. The above words must have sounded harsh and will I fear have given pain to many a tender heart which is conscious of the depth of its own love for the Redeemer, and would be shocked at the thought that anything had been neglected in his service, but has not the voice of such a heart returned answer to itself that what I have written is just?

Again, I have been told by some that they have been aware of the necessity of doing their best towards putting a stop to infidelity, and that they have been unceasing in their prayers for friends or husbands or relations who know not Christ, but that with prayers their efforts have ended. Now, there can be no one in the whole world who has had more signal proofs of the efficacy of prayer than the writer of these pages, but he would lie if he were to say that prayer was ever answered when it was only another name for idleness, a cloak for the avoidance of obvious duty. God is no helper of the indolent and the coward; if this were so, what need to work at all? Why not sit still, and trust in prayer for everything? No; to the women who have prayed, and prayed only, the answer is ready at hand, that work without prayer is bad, but prayer without work worse. Let them do their own utmost in the way of sowing, planting, and watering, and then let them pray to God that he will vouchsafe them the increase; but they can no more expect the increase to be of God's free gift without the toil of sowing than did the blessed Apostle St. Paul. If God did not convert the heathen for Paul and Apollos in answer to their prayers alone, how can we expect that he will convert the infidel for ourselves, unless we have first followed in the footsteps of the Apostles? The sin of infidelity will rest upon us and our children until we have done our best to shake it off; and this not timidly and disingenuously as those who fear for the result, but with the certainty that it is the infidel and not the Christian who need fear investigation, if the investigation only goes deep enough. Herein has lain our error, we have feared to allow the unbeliever to put forth all his strength lest it should prove stronger than we thought it was, when in truth the world would only have known the sooner of its weakness; and this shall now at last be abundantly shewn, for, as I said above, I will help no infidel by concealing his case; it shall appear in full, and as nearly in his own words as the limits at my disposal will allow. Out of his own mouth shall he be condemned, and yet, I trust, not condemned alone; but converted as I myself, and by the same irresistible chain of purest reason; one thing only is wanted on the part of the reader, it is this, the desire to attain truth regardless of past prejudices.

If an unbeliever has made up his mind that we must be wrong, without having heard our side, and if he presumes to neglect the most ordinary precaution against error—that of understanding the position of an opponent—I can do nothing with him or for him. No man can make another see, if the other persists in shutting his eyes and bandaging them: if it is a victory to be able to say that they cannot see the truth under these circumstances, the victory is with our opponents; but for those who can lay their hands upon their heart and say truly before God and man that they care nothing for the maintenance of their own opinions, but only that they may come to know the truth, for such I can do much. I can put the matter before them in so clear a light that they shall never doubt hereafter.

Never was there a time when such an exposition was wanted so much as now. The specious plausibilities of a pseudo-science have led hundreds of thousands into error; the misapplication of geology has ensnared a host of victims, and a still greater misapplication of natural history seems likely to devour those whom the perversion of geology has spared. Not that I have a word to say against true science: true science can never be an enemy of the Bible, which is the text-book of the science of the salvation of human souls as written by the great Creator and Redeemer of the soul itself, but the Enemy of Mankind is never idle, and no sooner does God vouchsafe to us any clearer illumination of his purposes and manner of working, than the Evil One sets himself to consider how he can turn the blessing into a curse; and by the all-wise dispensation of Providence he is allowed so much triumph as that he shall sift the wise from the foolish, the faithful from the traitors. God knoweth his own. Still there is no surer mark that one is among the number of those whom he hath chosen than the desire to bring all to share in the gracious promises which he has vouchsafed to those that will take advantage of them; and there are few more certain signs of reprobation than indifference as to the existence of unbelief, and faint-heartedness in trying to remove it. It is the duty of all those who love Christ to lead their brethren to love him also; but how can they hope to succeed in this until they understand the grounds on which he is rejected?

For there are grounds, insufficient ones, untenable ones, grounds which a little loving patience and, if I may be allowed the word, ingenuity, will shew to be utterly rotten; but as long as their rottenness is only to be asserted and not proved, so long will deluded people build upon them in fancied security. As yet the proof has never been made sufficiently clear. If displayed sufficiently for one age it has been necessary to do the work again for the next. As soon as the errors of one set of people have been made apparent, another set has arisen with fresh objections, or the old fallacies have reappeared in another shape. It is not too much to say that it has never yet been so clearly proved that Christ rose again from the dead, that a jury of educated Englishmen should be compelled to assent to it, even though they had never before heard of Christianity. This therefore it is my object to do once and for ever now.

It is not for me to pry into the motives of the Almighty, nor to inquire why it is that for nearly two thousand years the perfection of proof should never have been duly produced, but if I dare hazard an opinion I should say that such proof was never necessary until now, but that it has lain ready to be produced at a moment's notice on the arrival of the fitting time. In the early stages of the Church the vivâ voce testimony of the Apostles was still so near that its force was in no way spent; from those times until recently the universality of belief was such that proof was hardly needed; it is only for a hundred years or so (which in the sight of God are but as yesterday) that infidelity has made real progress. Then God raised his hand in wrath; revolution taught men to see the nature of unbelief and the world shrank back in horror; the time of fear passed by; unbelief has again raised itself; whereon we can see that other and even more fearful revolutions {82} are daily threatening. What country is safe? In what part of the world do not men feel an uneasy foreboding of the wrath which will surely come if they do not repent and turn unto the Lord their God? Go where we will we are conscious of that heaviness and oppression which is the precursor of the hurricane and the earthquake; none escape it: an all-pervading sense of rottenness and fearful waiting upon judgment is upon the hearts of all men. May it not be that this awe and silence have been ordained in order that the still small voice of the Lord may be the more clearly heard and welcomed as salvation? Is it not possible that the infinite mercy of God is determined to give mankind one last chance, before the day of that coming which no creature may abide? I dare not answer: yet I know well that the fire burneth within me, and that night and day I take no rest but am consumed until the work committed to me is done, that I may be clear from the blood of all men.

CHAPTER II

STRAUSS AND THE HALLUCINATION THEORY

It has been well established by Paley, and indeed has seldom been denied, that within a very few years of Christ's crucifixion a large number of people believed that he had risen from the dead. They believed that after having suffered actual death he rose to actual life, as a man who could eat and drink and talk, who could be seen and handled. Some who held this were near relations of Christ, some had known him intimately for a considerable time before his crucifixion, many must have known him well by sight, but all were unanimous in their assertion that they had seen him alive after he had been dead, and in consequence of this belief they adopted a new mode of life, abandoning in many cases every other earthly consideration save that of bearing witness to what they had known and seen. I have not thought it worth while to waste time and space by introducing actual proof of the above. This will be found in Paley's opening chapters, to which the reader is referred.

How then did this intensity of conviction come about? Differ as they might and did upon many of the questions arising out of the main fact which they taught, as to the fact itself they differed not in the least degree. In their own life-time and in that of those who could confute them their story gained the adherence of a very large and ever increasing number. If it could be shewn that the belief in Christ's reappearance did not arise until after the death of those who were said to have seen him, when actions and teachings might have been imputed to them which were not theirs, the case would then be different; but this cannot be done; there is nothing in history better established than that the men who said that they had seen Christ alive after he had been dead, were themselves the first to lay aside all else in order to maintain their assertion. If it could be maintained that they taught what they did in order to sanction laxity of morals, the case would again be changed. But this too is impossible. They taught what they did because of the intensity of their own conviction and from no other motive whatsoever.

What then can that thing have been which made these men so beyond all measure and one-mindedly certain? Were they thus before the Crucifixion? Far otherwise. Yet the men who fled in the hour of their master's peril betrayed no signs of flinching when their own was no less imminent. How came it that the cowardice and fretfulness of the Gospels should be transformed into the lion-hearted steadfastness of the Acts?

The Crucifixion had intervened. Yes, but surely something more than the Crucifixion. Can we believe that if their experience of Christ had ended with the Cross, the Apostles would have been in that state of mind which should compel them to leave all else for the sake of preaching what he had taught them? It is a hard thing for a man to change the scheme of his life; yet this is not a case of one man but of many, who became changed as if struck with an enchanter's wand, and who, though many, were as one in the vehemence with which they protested that their master had reappeared to them alive. Their converse with Christ did not probably last above a year or two, and was interrupted by frequent absence. If Christ had died once and for all upon the Cross, Christianity must have died with him; but it did not die; nay, it did not begin to live with full energy until after its founder had been crucified. We must ask again, what could that thing have been which turned these querulous and faint-hearted followers into the most earnest and successful body of propagandists which the world has ever seen, if it was not that which they said it was—namely, that Christ had reappeared to them alive after they had themselves known him to be dead? This would account for the change in them, but is there anything else that will?

They had such ample opportunities of knowing the truth that the supposition of mistake is fraught with the greatest difficulties; they gave such guarantees of sincerity as that none have given greater; their unanimity is perfect; there is not the faintest trace of any difference of opinion amongst them as to the main fact of the Resurrection. These are things which never have been and never can be denied, but if they do not form strong primâ facie ground for believing in the truth and actuality of Christ's Resurrection, what is there which will amount to a primâ facie case for anything whatever?

Nevertheless the matter does not rest here. While there exists the faintest possibility of mistake we may be sure that we shall deal most wisely by examining its character and value. Let us inquire therefore whether there are any circumstances which seem to indicate that the early Christians might have been mistaken, and been firmly persuaded that they had seen Christ alive, although in point of fact they had not really seen him? Men have been very positive and very sincere about things wherein we should have conceived mistake impossible, and yet they have been utterly mistaken. A strong predisposition, a rare coincidence, an unwonted natural phenomenon, a hundred other causes, may turn sound judgments awry, and we dare not assume forthwith that the first disciples of Christ were superior to influences which have misled many who have had better chances of withstanding them. Visions and hallucinations are not uncommon even now. How easily belief in a supernatural occurrence obtains among the peasantry of Italy, Ireland, Belgium, France, and Spain; and how much more easily would it do so among Jews in the days of Christ, when belief in supernatural interferences with this world's economy was, so to speak, omnipresent. Means of communication, that is to say of verification, were few, and the tone of men's minds as regards accuracy of all kinds was utterly different from that of our own; science existed not even in name as the thing we now mean by it; few could read and fewer write, so that a story could seldom be confined to its original limits; error, therefore, had much chance and truth little as compared with our own times. What more is needed to make us feel how possible it was for the purest and most honest of men to become parents of all fallacy?

Strauss believes this to have been the case. He supposes that the earliest Christians were under hallucination when they thought that they had seen Christ alive after his Crucifixion; in other words, that they never saw him at all, but only thought that they had done so. He does not imagine that they conceived this idea at once, but that it grew up gradually in the course of a few years, and that those who came under its influence antedated it unconsciously afterwards. He appears to believe that within a few months of the Crucifixion, and in consequence of some unexplained combination of internal and external causes, some one of the Apostles came to be impressed with the notion that he had seen Christ alive; the impression, however made, was exceedingly strong, and was communicated as soon as might be to some other or others of the Apostles: the idea was welcome—as giving life to a hope which had been fondly cherished; each inflamed the imagination of the other, until the original basis of the conception slipped unconsciously from recollection, while the intensity of the conviction itself became stronger and stronger the more often the story was repeated. Strauss supposes that on seeing the firm conviction of two or three who had hitherto been leaders among them, the other Apostles took heart, and that thus the body grew together again perhaps within a twelve-month of the Crucifixion. According to him, the idea of the Resurrection having been once started, and having once taken root, the soil was so congenial that it grew apace; the rest of the Apostles, perhaps assembled together in a high state of mental enthusiasm and excitement, conceived that they saw Christ enter the room in which they were sitting and afford some manifest proof of life and identity; or some one else may have enlarged a less extraordinary story to these dimensions, so that in a short time it passed current everywhere (there have been instances of delusions quite as extraordinary gaining a foothold among men whose sincerity is not to be disputed), and finally they conceived that these appearances of their master had commenced a few months—and what is a few months?—earlier than they actually had, so that the first appearance was soon looked upon as having been vouchsafed within three days of the Crucifixion.

The above is not in Strauss's words, but it is a careful résumé of what I gather to be his conception of the origin of the belief in the Resurrection of Christ. The belief, and the intensity of the belief, need explanation; the supernatural explanation, as we should ourselves readily admit, cannot be accepted unless all others are found wanting; he therefore, if I understand him rightly, puts forward the above as being a reasonable and natural solution of the difficulty—the only solution which does not fail upon examination, and therefore the one which should be accepted. It is founded upon the affection which the Apostles had borne towards their master, and their unwillingness to give up their hope that they had been chosen, as the favoured lieutenants of the promised Messiah.

No man would be willing to give up such hope easily; all men would readily welcome its renewal; it was easy in the then intellectual condition of Palestine for hallucination to originate, and still easier for it to spread; the story touched the hearts of men too nearly to render its propagation difficult. Men and women like believing in the marvellous, for it brings the chance of good fortune nearer to their own doors; but how much more so when they are themselves closely connected with the central figure of the marvel, and when it appears to give a clue to the solution of that mystery which all would pry into if they could—our future after death? There can be no great cause for wonder that an hallucination which arose under such conditions as these should have gained ground and conquered all opposition, even though its origin may be traced to the brain of but a single person.

He would be a bold man who should say that this was impossible; nevertheless it cannot be accepted. For, in the first place, we collect most certainly from the Gospel records that the Apostles were not a compact and devoted body of adherents at the time of the Crucifixion; yet it is hard to see how Strauss's hallucination theory can be accepted, unless this was the case. If Strauss believed the earliest followers of Christ to have been already immovably fixed in their belief that he was the Son of God—the promised Messiah, of whom they were themselves the especially chosen ministers—if he considered that they believed in their master as the worker of innumerable miracles which they had themselves witnessed; as one whom they had seen raise others from death to life, and whom, therefore, death could not be expected to control—if he held the followers of Christ to have been in this frame of mind at the time of the Crucifixion, it might be intelligible that he should suppose the strength of their faith to have engendered an imaginary reappearance in order to save them from the conclusion that their hopes had been without foundation; that, in point of fact, they should have accepted a new delusion in order to prop up an old one; but we know very well that Strauss does not accept this position. He denies that the Apostles had seen any miracles; independently therefore of the many and unmistakable traces of their having been but partial and wavering adherents, which have made it a matter of common belief among those who have studied the New Testament that the faith of the Apostles was unsteadfast before the Crucifixion, he must have other and stronger reasons for thinking that this was so, inasmuch as he does not look upon them as men who had seen our Lord raise any one from the dead, nor restore the eyes of the blind.

According to him, they may have seen Christ exercise unusual power over the insane, and temporary alleviations of sickness, due perhaps to mental excitement, may have taken place in their presence and passed for miracles; he would doubt how far they had even seen this much, for he would insist on many passages in the Gospels which would point in the direction of our Lord's never having professed to work a single miracle; but even though he granted that they had seen certain extraordinary cases of healing, there is no amount of testimony which would for a moment satisfy him of their having seen more. We see the Apostles as men who before the Crucifixion had seen Lazarus raised from death to life after the corruption of the grave had begun its work, and who had seen sight given to one that had been born sightless; as men who had seen miracle after miracle, with every loophole for escape from a belief in the miraculous carefully excluded; who had seen their master walking upon the sea, and bidding the winds be still; our difficulty therefore is to understand the incredulity of the Apostles as displayed abundantly in the Gospels; but Strauss can have none such; for he must see them as men over whom the influence of their master had been purely personal, and due to nothing more than to a strength and beauty of character which his followers very imperfectly understood. He does not believe that Lazarus was raised at all, or that the man who had been born blind ever existed; he considers the fourth gospel, which alone records these events, to be the work of a later age, and not to be depended on for facts, save here and there; certainly not where the facts recorded are miraculous. He must therefore be even more ready than we are to admit that the faith of the Apostles was weak before the Crucifixion; but whether he is or not, we have it on the highest authority that their faith was not strong enough to maintain them at the very first approach of danger, nor to have given them any hope whatever that our Lord should rise again; whereas for Strauss's theory to hold good, it must already have been in a white heat of enthusiasm.

But even granting that this was so—in the face of all the evidence we can reach—men so honest and sincere as the Apostles proved themselves to be, would have taken other ground than the assertion that their master had reappeared to them alive, unless some very extraordinary occurrences had led them to believe that they had indeed seen him. If their faith was glowing and intense at the time of the Crucifixion—so intense that they believed in Christ as much, or nearly as much, after the Crucifixion as before it (and unless this were so the hallucinations could never have arisen at all, or at any rate could never have been so unanimously accepted)—it would have been so intense as to stand in no need of a reappearance. In this case, if they had found that their master did not return to them, the Apostles would probably have accepted the position that he had, contrary to their expectation, been put to a violent death; they would, perhaps, have come sooner or later to the conclusion that he was immediately on death received into Heaven, and was sitting on the right hand of God; while some extraordinary dream might have been construed into a revelation of the fact with the manner of its occurrence, and been soon generally believed; but the idea of our Lord's return to earth in a gross material body whereon the wounds were still unhealed, was perhaps the last thing that would have suggested itself to them by way of hallucination. If their faith had been great enough, and their spirits high enough to have allowed hallucination to originate at all, their imagination would have presented them at once with a glorious throne, and the splendours of the highest Heaven as appearing through the opened firmament; it would not surely have rested satisfied with a man whose hands and side were wounded, and who could eat of a piece of broiled fish and of an honeycomb. A fabric so utterly baseless as the reappearances of our Lord (on the supposition of their being unhistoric) would have been built of gaudier materials. To repeat, it seems impossible that the Apostles should have attempted to connect their hallucinations circumstantially and historically with the events which had immediately preceded them. Hallucination would have been conscious of a hiatus and not have tried to bridge it over. It would not have developed the idea of our Lord's return to this grovelling and unworthy earth prior to his assumption into glory, unless those who were under its influence had either seen other resurrections from the dead—in which case there is no difficulty attaching to the Resurrection of our Lord himself—or been forced into believing it by the evidence of their own senses; this, on the supposition that the devotion of the first disciples was intense before the Crucifixion; but if, on the other hand, they were at that time anything but steadfast, as both a priori and a posteriori evidence would seem to indicate, if they were few and wavering, and if what little faith they had was shaken to its foundations and apparently at an end for ever with the death of Christ, it becomes indeed difficult to see how the idea of his return to earth alive could have ever struck even a single one of them, much less that hallucinations which could have had no origin but in the disordered brain of some one member of the Apostolic body, should in a short time have been accepted by all as by one man without a shadow of dissension, and been strong enough to convert them, as was said above, into the most earnest and successful body of propagandists that the world has ever seen.

Truly this is not too much to say of them; and yet we are asked to believe that this faith, so intensely energetic, grew out of one which can hardly be called a faith at all, in consequence of day-dreams whose existence presupposes a faith hardly if any less intense than that which it is supposed to have engendered. Are we not warranted in asserting that a movement which is confined to a few wavering followers, and which receives any very decisive check, which scatters and demoralises the few who have already joined it, will be absolutely sure to die a speedy natural death unless something utterly strange and new occurs to give it a fresh impetus? Such a resuscitating influence would have been given to the Christian religion by the reappearance of Christ alive. This would meet the requirements of the case, for we can all feel that if we had already half believed in some gifted friend as a messenger from God, and if we had seen that friend put to death before our eyes, and yet found that the grave had no power over him, but that he could burst its bonds and show himself to us again unmistakably alive, we should from that moment yield ourselves absolutely his; but our faith would die with him unless it had been utter before his death.

The devotion of the Apostles is explained by their belief in the Resurrection, but their belief in the Resurrection is not explained by a supposed hallucination; for their minds were not in that state in which alone such a delusion could establish itself firmly, and unless it were established firmly by the most apparently irrefragable evidence of many persons, it would have had no living energy. How an hallucination could occur in the requisite strength to the requisite number of people is neither explained nor explicable, except upon the supposition that the Apostles were in a very different frame of mind at the time of Christ's Crucifixion from that which all the evidence we can get would seem to indicate. If Strauss had first made this point clear we could follow him. But he has not done so.

Strauss says, the conception that Christ's body had been reawakened and changed, "a double miracle, exceeding far what had occurred in the case of Enoch and Elijah, could only be credible to one who saw in him a prophet far superior to them"—i.e., to one who notwithstanding his death was persuaded that he was the Messiah: "this conviction" (that a double miracle had been performed) "was the first to which the Apostles had to attain in the days of their humiliation after the Crucifixion." Yes—but how were they to attain to it, being now utterly broken down and disillusioned? Strauss admits that before they could have come to hold what he supposes them to have held, they must have seen in Christ even after his Crucifixion a prophet far greater than either Moses or Elias; whereas in point of fact it is very doubtful whether they ever believed this much of their master even before the Crucifixion, and hardly questionable that after it they disbelieved in him almost entirely, until he shewed himself to them alive.

Is it possible that from the dead embers of so weak a faith, so vast a conflagration should have been kindled?

I submit, therefore, that independently of any direct evidence as to the when and where of Christ's reappearances, the fact that the Apostles before the Crucifixion were irresolute, and after it unspeakably resolute, affords strong ground for believing that they must have seen something, or come to know something, which to their minds was utterly overwhelming in its convincing power: when we find the earliest and most trustworthy records unanimously asserting that that something was the reappearance of Christ alive, we feel that such a reappearance was an adequate cause for the result actually produced; and when we think over the condition of mind which both probability and evidence assign to the Apostles, we also feel that no other circumstance would have been adequate, nor even this unless the proof had been such as none could reasonably escape from.

Again, Strauss's supposition that the Apostles antedated their hallucinations suggests no less difficulty. Suppose that, after all, Strauss is right, and that there was no actual reappearance; whatever it was that led the Apostles to believe in such reappearance must have been, judging by its effect, intense and memorable: it must have been as a shock obliterating everything save the memory of itself and the things connected with it: the time and manner of such a shock could never have been forgotten, nor misplaced without deliberate intention to deceive, and no one will impute any such intention to the Apostles.

It may be said that if they were capable of believing in the reality of their visions they would be also capable of antedating them; this is true; but the double supposition of self-delusion, first in seeing the visions at all, and then in unconsciously antedating them, reduces the Apostles to such an exceedingly low level of intelligence and trustworthiness, that no good and permanent work could come from such persons; the men who could be weak enough, and crazed enough, if the reader will pardon the expression, to do as Strauss suggests, could never have carried their work through in the way they did. Such men would have wrecked their undertaking a hundred times over in the perils which awaited it upon every side; they would have become victims of their own fancies and desires, with little or no other grounds than these for any opinions they might hold or teach: from such a condition of mind they must have gone on to one still worse; and their tenets would have perished with them, if not sooner.

Again, as regards this antedating; unless the visions happened at once, it is inconceivable that they should have happened at all. Strauss believes that the disciples fled in their first terror to their homes: that when there, "outside the range to which the power of the enemies and murderers of their master extended, the spell of terror and consternation which had been laid upon their minds gave way," and that under the circumstances a reaction up to the point at which they might have visions of Christ is capable of explanation. The answer to this is that it is indeed likely that the spell of terror would give way when they found themselves safe at home, but that it is not at all likely that any reaction would take place in favour of one to whom their allegiance had never been thorough, and whom they supposed to have met with a violent and accursed end. It might be easy to imagine such a reaction if we did not also attempt to imagine the circumstances that must have preceded it; the moment we try to do this, we find it to be an impossibility. If once the Apostles had been dispersed, and had returned home to their former avocations without having seen or heard anything of their master's return to earth, all their expectations would have been ended; they would have remained peaceable fishermen for the rest of their lives, and been cured once and for ever of their enthusiasm.

Can we believe that the disciples, returning to Galilee in fear, and bereaved of that master mind which had kept them from falling out with one another, would have remained a united and enthusiastic body? Strauss admits that their enthusiasm was for the time ended. Is it then likely that they would have remained in any sense united, or is it not much more likely that they would have shunned each other and disliked allusions to the past? What but Christ's actual reappearance could rekindle this dead enthusiasm, and fan it to such a burning heat? Suppose that one or two disciples recovered faith and courage, the majority would never do so. If Christ himself with the magic of his presence could not weld them into a devoted and harmonious company, would the rumour arising at a later time that some one had seen him after death, be acceptable enough to make the others believe that they too had actually seen and handled him? Perhaps—if the rumour was believed. But would it have been believed? Or at any rate have been believed so utterly?

We cannot think it. For the belief and assertion are absolutely without trace of dissent within the Christian body, and that body was in the first instance composed entirely of the very persons who had known and followed Christ before the Crucifixion. If some of the original twelve had remained aloof and disputed the reappearances of Christ, is it possible that no trace of such dissension should appear in the Epistles of St. Paul? Paul differed widely enough from those who were Apostles before him, and his language concerning them is occasionally that of ill-concealed contempt and hatred rather than of affection; but is there a word or hint which would seem to indicate that a single one of those who had the best means of knowing doubted the Resurrection? There is nothing of the kind; on the contrary, whatever we find is such as to make us feel perfectly sure that none of them did doubt it. Is it then possible that this unanimity should have sprung from the original hallucinations of a small minority? True—it is plain from the Epistle to the Corinthians that there were some of Paul's contemporaries who denied the Resurrection. But who were they? We should expect that many among the more educated Gentile converts would throw doubt upon so stupendous a miracle, but is there anything which would point in the direction of these doubts having been held within the original body of those who said that they had seen Christ alive? By the eleven, or by the five hundred who saw him at once? There is not one single syllable. Those who heard the story second-hand would doubtless some of them attempt to explain away its miraculous character, but if it had been founded on hallucination it is not from these alone that the doubts would have come.

Something is imperatively demanded in order to account for the intensity of conviction manifested by the earliest Christians shortly after the Crucifixion; for until that time they were far from being firmly convinced, and the Crucifixion was the very last thing to have convinced them. Given (to speak of our Lord as he must probably appear to Strauss) an unusually gifted teacher of a noble and beautiful character: given also, a small body of adherents who were inclined to adopt him as their master and to regard him as the coming liberator, but who were nevertheless far from settled in their conviction: given such a man and such followers: the teacher is put to a shameful death about two years after they had first known him, and the followers forsake him instantly: surely without his reappearing in some way upon the scene they would have concluded that their doubts had been right and their hopes without foundation: but if he reappeared, their faith would, for the first time, become intense, all-absorbing. Surely also they might be trusted to know whether they had really seen their master return to them or not, and not to sacrifice themselves in every way, and spend their whole lives in bearing testimony to pure hallucination?

There is one other point on which a few words will be necessary, before we proceed to the arguments in favour of the objective character of Christ's Resurrection as derivable from the conversion and testimony of St. Paul. It is this. Strauss and those who agree with him will perhaps maintain that the

Apostles were in truth wholly devoted to Christ before the Crucifixion, but that the Evangelists have represented them as being only half-hearted, in order to heighten the effect of their subsequent intense devotion. But this looks like falling into the very error which Rationalists condemn most loudly when it comes from so-called orthodox writers. They complain, and with too much justice, that our apologists have made "anything out of anything." Yet if the Apostles were not unsteadfast, and did not desert their master in his hour of peril, and if all the accounts of Christ's reappearances are the creations of disordered fancy, we may as well at once declare the Evangelists to be worthless as historians, and had better give up all attempt at the construction of history with their assistance. We cannot take whatever we wish, and leave whatever we wish, and alter whatever we wish. If we admit that upon the whole the Gospel writings or at any rate the first three Gospels, contain a considerable amount of historic matter, we should also arrive at some general principles by which we will consistently abide in separating the historic from the unhistoric. We cannot deal with them arbitrarily, accepting whatever fits in with our fancies, and rejecting whatever is at variance with them.

Now can it be maintained that the Evangelists would be so likely to overrate the half-heartedness of the Apostles, that we should look with suspicion upon the many and very plain indications of their having been only half-hearted? Certainly not. If there was any likelihood of a tendency one way or the other it would be in the direction of overrating their faith. Would not the unbelief of the Apostles in the face of all the recorded miracles be a most damaging thing in the eyes of the unconverted? Would not the Apostles themselves, after they were once firmly convinced, be inclined to think that they had from the first believed more firmly than they really had done? This at least would be in accordance with the natural promptings of human instinct: we are all of us apt to be wise after the event, and are far more prone to dwell upon things which seem to give some colour to a pretence of prescience, than upon those which force from us a confession of our own stupidity. It might seem a damaging thing that the Apostles should have doubted as much as long as they clearly did; would then the Evangelists go out of their way to introduce more signs of hesitation? Would any one suggest that the signs of doubt and wavering had been overrated, unless there were some theory or other to be supported, in order to account for which this overrating was necessary? Would the opinion that the want of faith had been exaggerated arise prior to the formation of a theory, or subsequently? This is the fairest test; let the reader apply it for himself.

On the other hand, there are many reasons which should incline us to believe that, before the Resurrection, the Apostles were less convinced than is generally supposed, but it would be dangerous to depart either to the right hand or to the left of that which we find actually recorded, namely, that in the main the Apostles were prepared to accept Christ before the Crucifixion, but that they were by no means resolute and devoted followers. I submit that this is a fair rendering of the spirit of what we find in the Gospels. It is just because Strauss has chosen to depart from it that he has found himself involved in the maze of self-contradiction through which we have been trying to follow him. There is no position so absurd that it cannot be easily made to look plausible, if the strictly scientific method of investigation is once departed from.

But if I had been in Strauss's place, and had wished to make out a case against Christianity without much heed of facts, I should not have done it by a theory of hallucinations. A much prettier, more novel and more sensational opening for such an attempt is afforded by an attack upon the Crucifixion itself. A very neat theory might be made, that there may have been some disturbance at one of the Jewish passovers, during which some persons were crucified as an example by the Romans: that during this time Christ happened to be missing; that he reappeared, and finally departed, whither, no man can say: that the Apostles, after his last disappearance, remembering that he had been absent during the tumult, little by little worked themselves up into the belief that on his reappearance they had seen wounds upon him, and that the details of the Crucifixion were afterwards revealed in a vision to some favoured believer, until in the course of a few years the narrative assumed its present shape: that then the reappearance of Christ was denied among the Jews, while the Crucifixion as attaching disgrace to him was not disputed, and that it thus became so generally accepted as to find its way into Pliny and Josephus. This tissue of absurdity may serve as an example of what the unlicensed indulgence of theory might lead to; but truly it would be found quite as easy of belief as that the early Christian faith in the Resurrection was due to hallucination only.

Considering, then, that Christianity was not crushed but overran the most civilised portions of the world; that St. Paul was undoubtedly early told, in such a manner as for him to be thoroughly convinced of the fact, that on some few but sufficient occasions Christ was seen alive after he had been crucified; that the general belief in the reappearance of our Lord was so strong that those who had the best means of judging gave up all else to preach it, with a unanimity and singleness of purpose which is irreconcilable with hallucination; that all our records most definitely insist upon this belief and that there is no trace of its ever having been disputed among the Jewish Christians, it seems hard to see how we can escape from admitting that Jesus Christ was crucified, dead, and buried, and yet that he was verily and indeed seen alive again by those who expected nothing less, but who, being once convinced, turned the whole world after them.

It is now incumbent upon us to examine the testimony of St. Paul, to which I would propose to devote a separate chapter.

CHAPTER III

THE CHARACTER AND CONVERSION OF ST PAUL

Setting aside for the present the story of St. Paul's conversion as given in the Acts of the Apostles—for I am bound to admit that there are circumstances in connection with that account which throw doubt upon its historical accuracy—and looking at the broad facts only, we are struck at once with the following obvious reflection, namely, that Paul was an able man, a cultivated man, and a bitter opponent of Christianity; but that in spite of the strength of his original prejudices, he came to see what he thought convincing reasons for going over to the camp of his enemies. He went over, and with the result we are all familiar.

Now even supposing that the miraculous account of Paul's conversion is entirely devoid of foundation, or again, as I believe myself, that the story given in the Acts is not correctly placed, but refers to the vision alluded to by Paul himself (I. Cor. xv.), and to events which happened, not coincidently with his conversion, but some years after it—does not the importance of the conversion itself rather gain than lose in consequence? A charge of unimportant inaccuracy may be thus sustained against one who wrote in a most inaccurate age; but what is this in comparison with the testimony borne to the strength of the Christian evidences by the supposition that of their own weight alone, and without miraculous assistance, they succeeded in convincing the most bitter, and at the same time the ablest, of their opponents? This is very pregnant. No man likes to abandon the side which he has once taken. The spectacle of a man committing himself deeply to his original party, changing without rhyme or reason, and then remaining for the rest of his life the most devoted and courageous adherent of all that he had opposed, without a single human inducement to make him do so, is one which has never been witnessed since man was man. When men who have been committed deeply and spontaneously to one cause, leave it for another, they do so either because facts have come to their knowledge which are new to them and which they cannot resist, or because their temporal interests urge them, or from caprice: but if they change from caprice in important matters and after many pledges given, they will change from caprice again: they will not remain for twenty-five or thirty years without changing a jot of their capriciously formed opinions. We are therefore warranted in assuming that St. Paul's conversion to Christianity was not dictated by caprice: it was not dictated by self-interest: it must therefore have sprung from the weight of certain new facts which overbore all the resistance which he could make to them.

What then could these facts have been?

Paul's conduct as a Jew was logical and consistent: he did what any seriously-minded man who had been strictly brought up would have done in his situation. Instead of half believing what he had been taught, he believed it wholly. Christianity was cutting at the root of what was in his day accepted as fundamental: it was therefore perfectly natural that he should set himself to attack it. There is nothing against him in this beyond the fact of his having done it, as far as we can see, with much cruelty. Yet though cruel, he was cruel from the best of motives—the stamping out of an error which was harmful to the service of God; and cruelty was not then what it is now: the age was not sensitive and the lot of all was harder. From the first he proved himself to be a man of great strength of character, and like many such, deeply convinced of the soundness of his opinions, and deeply impressed with the belief that nothing could be good which did not also commend itself as good to him. He tested the truth of his earlier convictions not by external standards, but by the internal standard of their own strength and purity—a fearful error which but for God's mercy towards him would have made him no less wicked than well-intentioned.

Even after having been convinced by a weight of evidence which no prejudice could resist, and after thus attaining to a higher conception of right and truth and goodness than was possible to him as a Jew, there remained not a few traces of the old character. Opposition beyond certain limits was a thing which to the end of his life he could not brook. It is not too much to say that he regarded the other Apostles— and was regarded by them—with suspicion and dislike; even if an angel from Heaven had preached any other doctrine than what Paul preached, the angel was to be accursed (Gal. i., 8), and it is not probable that he regarded his fellow Apostles as teaching the same doctrine as himself, or that he would have allowed them greater licence than an angel. It is plain from his undoubted Epistles to the Corinthians and Galatians that the other Apostles, no less than his converts, exceedingly well knew that he was not a man to be trifled with. If the arm of the law had been as much on his side after his conversion as before it, it would have gone hardly with dissenters; they would have been treated with politic tenderness the moment that they yielded, but woe betide them if they presumed on having any very decided opinions of their own.

On the other hand, his sagacity is beyond dispute; it is certain that his perception of what the Gentile converts could and could not bear was the main proximate cause of the spread of Christianity. He prevented it from becoming a mere Jewish sect, and it has been well said that but for him the Jews would now be Christians, and the Gentiles unbelievers. Who can doubt his tact and forbearance, where matters not essential were concerned? His strength in not yielding a fraction upon vital points was matched only by his suppleness and conciliatory bearing upon all others. To use his own words, he did indeed become "all things to all men" if by any means he could gain some, and the probability is that he pushed this principle to its extreme (see Acts xxi., 20–26).

Now when we see a man so strong and yet so yielding—the writer moreover of letters which shew an intellect at once very vigorous and very subtle (not to say more of them), and when we know that there was no amount of hardship, pain, and indignity, which he did not bear and count as gain in the service of Jesus Christ; when we also remember that he continued thus for all the known years of his life after his conversion, can we think that that conversion could have been the result of anything even approaching to caprice? Or again, is it likely that it could have been due to contact with the hallucinations of his despised and hated enemies? Paul the Christian appears to be the same sort of man in most respects as Paul the Jew, yet can we imagine Paul the Christian as being converted from Christianity to some other creed, by the infection of hallucinations? On the contrary, no man would more quickly have come to the bottom of them, and assigned them to diabolical agency. What then can that thing have been, which wrenched the strong and able man from all that had the greatest hold upon him, and fixed him for the rest of his life as the most self-sacrificing champion of Christianity? In answer to this question we might say, that it is of no great importance how the change was made, inasmuch as the fact of its having been made at all is sufficiently pregnant. Nevertheless it will be interesting to follow Strauss in his remarks upon the account given in the Acts, and I am bound to add that I think he has made out his case. Strange! that he should have failed to see that the evidences in support of the Resurrection are incalculably strengthened by his having done so. How short-sighted is mere ingenuity! And how weak and cowardly are they who shut their eyes to facts because they happen to come from an opponent!

Strauss, however, writes as follows:—"That we are not bound to the individual features of the account in the Acts is shewn by comparing it with the substance of the statement twice repeated in the language of Paul himself: for there we find that the author's own account is not accurate, and that he attributed no importance to a few variations more or less. Not only is it said on one occasion that the attendants stood dumb-foundered: on another that they fell with Paul to the ground; on one occasion that they heard the voice but saw no one; on another that they saw the light but did not hear the voice of him who spoke with Paul: but also the speech of Jesus himself, in the third repetition, gets the well known addition about "kicking against the pricks," to say nothing of the fact that the appointment to the Apostleship of the Gentiles, which according to the two earlier accounts was made partly by Ananias, partly on the occasion of a subsequent vision in the Temple at Jerusalem, is in this last account incorporated in the speech of Jesus. There is no occasion to derive the three accounts of this occurrence in the Acts from different sources, and even in this case one must suppose that the author of the Acts must have remarked and reconciled the discrepancies; that he did not do so, or rather that without following his own earlier narrative he repeated it in an arbitrary form, proves to us how careless the New Testament writers are about details of this kind, important as they are to one who strives after strict historical accuracy.

"But even if the author of the Acts had gone more accurately to work, still he was not an eye witness, scarcely even a writer who took the history from the narrative of an eye witness. Even if we consider the person who in different places comprehends himself and the Apostle Paul under the word 'we' or 'us' to have been the composer of the whole work, that person was not on the occasion of the occurrence before Damascus as yet in the company of the Apostle. Into this he did not enter until much later, in the Troad, on the Apostle's second missionary journey (Acts xvi., 10). But that hypothesis with regard to the author of the Acts of the Apostles is, moreover, as we have seen above, erroneous. He only worked up into different passages of his composition the memoranda of a temporary companion of the Apostle about the journeys performed in his company, and we are therefore not justified in considering the narrator to have been an eye witness in those passages and sections in which the 'we' is wanting. Now among these is found the very section in which appear the two accounts of his conversion which Paul gives, first, to the Jewish people in Jerusalem, secondly, to Agrippa and Festus in Cæsarea. The last occasion on which the 'we' was found was xxi., 18, that of the visit of Paul to James, and it does not appear again until xxvii., 1, when the subject is the Apostle's embarkation for Italy. Nothing therefore compels us to assume that we have in the reports of these speeches the account of any one who had been a party to the hearing of them, and, in them, Paul's own narrative of the occurrences that took place on his conversion."

The belief in the verbal inspiration of the Scriptures having been long given up by all who have considered the awful consequences which it entails, the Bible records have been opened to modern criticism:—the result has been that their general accuracy is amply proved, while at the same time the writers must be admitted to have fallen in with the feelings and customs of their own times, and must accordingly be allowed to have been occasionally guilty of what would in our own age be called inaccuracies. There is no dependence to be placed on the verbal, or indeed the substantial, accuracy of any ancient speeches, except those which we know to have been reported verbatim, they were (as with the Herodotean and Thucydidean speeches) in most cases the invention of the historian himself, as being what seemed most appropriate to be said by one in the position of the speaker. Reporting was a rare art among the ancients, and was confined to a few great centres of intellectual activity; accuracy, moreover, was not held to be of the same importance as at the present day. Yet without accurate reporting a speech perishes as soon as it is uttered, except in so far as it lives in the actions of those who hear it. Even a hundred years ago the invention of speeches was considered a matter of course, as in the well-known case of Dr. Johnson, than whom none could be more conscientious, and—according to his lights—accurate. I may perhaps be pardoned for quoting the passage in full from Boswell, who gives it on the authority of Mr. John Nichols; the italics are mine. "He said that the Parliamentary debates were the only part of his writings which then gave him any compunction: but that at the time he wrote them he had no conception that he was imposing upon the world, though they were frequently written from very slender materials, and often from none at all—the mere coinage of his own imagination. He never wrote any part of his works with equal velocity." (Boswell's Life of Johnson, chap. lxxxii.)

This is an extreme case, yet there can be no question about its truth. It is only one among the very many examples which could be adduced in order to shew that the appreciation of the value of accuracy is a thing of modern date only—a thing which we owe mainly to the chemical and mechanical sciences, wherein the inestimable difference between precision and inaccuracy became most speedily apparent. If the reader will pardon an apparent digression, I would remark that that sort of care is wanted on behalf of Christianity with which a cashier in a bank counts out the money that he tenders—counting it and recounting it as though he could never be sure enough before he allowed it to leave his hands. This caution would have saved the wasting of many lives, and the breaking of many hearts.

We, on the other hand, however reckless we may be ourselves, are in the habit of assuming that any historian whom we may have occasion to consult, and on whose testimony we would fain rely, must have himself weighed and re-weighed his words as the cashier his money; an error which arises from want of that sympathy which should make us bear constantly in mind what lights men had, under what influences they wrote, and what we should ourselves have done had we been so placed as they. But if any will maintain that though the general run of ancient speeches were, as those supposed to have been reported by Johnson, pure invention, yet that it is not likely that one reporting the words of Almighty God should have failed to feel the awful responsibility of his position, we can only answer that the writer of the Acts did most indisputably so fail, as is shewn by the various reports of those words which he has himself given: if he could in the innocency of his heart do this, and at one time report the Almighty as saying this, and at another that, as though, more or less, this or that were a matter of no moment, what certainty can we have concerning such a man that inaccuracy shall not elsewhere be found in him? None. He is a warped mirror which will distort every object that it reflects.

It follows, then, that from the Acts of the Apostles we have no data for arriving at any conclusion as to the manner of Paul's change of faith, nor the circumstances connected with it. To us the accounts there given should be simply non-existent; but this is not easy, for we have heard them too often and from too early an age to be able to escape their influence; yet we must assuredly ignore them if we are anxious to arrive at truth. We cannot let the story told in the Acts enter into any judgement which we may form concerning Paul's character. The desire to represent him as having been converted by miracle was very natural. He himself tells us that he saw visions, and received his apostleship by revelation—not necessarily at the time of, or immediately after, his conversion, but still at some period or other in his life; it would be the most natural thing in the world for the writer of the Acts to connect some version of one of these visions with the conversion itself: the dramatic effect would be heightened by making the change, while the change itself would be utterly unimportant in the eyes of such a writer; be this however as it may, we are only now concerned with the fact that we know nothing about Paul's conversion from the Acts of the Apostles, which should make us believe that that conversion was wrought in him by any other means, than by such an irresistible pressure of evidence as no sane person could withstand.

From the Apostle's own writings we can glean nothing about his conversion which would point in the direction of its having been sudden or miraculous. It is true that in the Epistle to the Galatians he says, "After it had pleased God to reveal his Son in me," but this expression does not preclude the supposition that his conversion may have been led up to by a gradual process, the culmination of which (if that) he alone regarded as miraculous. Thus we are forced to admit that we know nothing from any source concerning the manner and circumstances of St. Paul's change from Judaism to Christianity, and we can only conclude therefore that he changed because he found the weight of the evidence to be greater than he could resist. And this, as we have seen, is an exceedingly telling fact. The probability is, that coming much into contact with Christians through his persecution of them, and submitting them to the severest questioning, he found that they were in all respects sober plainspoken men, that their conviction was intense, their story coherent, and the doctrines which they had received simple and ennobling; that these results of many inquisitions were so unvarying that he found conviction stealing gradually upon him against his will; common honesty compelled him to inquire further; the answers pointed invariably in one direction only; until at length he found himself utterly unable to resist the weight of evidence which he had collected, and resolved, perhaps at the last suddenly, to yield himself a convert to Christianity.

Strauss says that, "in the presence of the believers in Jesus," the conviction that he was a false teacher—an impostor—"must have become every day more doubtful to him. They considered it not only publicly honourable to be as convinced of his Resurrection as they were of their own life—but they shewed also a state of mind, a quiet peace, a tranquil cheerfulness, even under suffering, which put to shame the restless and joyless zeal of their persecutor. Could he have been a false teacher who had adherents such as these? Could that have been a false pretence which gave such rest and security? on the one hand, he saw the new sect, in spite of all persecutions, nay, in consequence of them, extending their influence wider and wider round them; on the other, as their persecutor, he felt that inward tranquillity growing less and less which he could observe in so many ways in the persecuted. We cannot therefore be surprised if in hours of inward despondency and unhappiness he put to himself the question, 'Who after all is right, thou, or the crucified Galilean about whom these men are so enthusiastic?' And when he had got as far as this, the result, with his bodily and mental characteristics, naturally followed in an ecstasy in which the very same Christ whom up to this time he had so passionately persecuted, appeared to him in all the glory of which his adherents spoke so much, shewed him the perversity and folly of his conduct, and called him to come over to his service."

The above comes simply to this, that Paul in his constant contact with Christians found that they had more to say for themselves than he could answer, and should, one would have thought, have suggested to Strauss what he supposes to have occurred to Paul, namely, that it was not likely that these men had made a mistake in thinking that they had seen Christ alive after his Crucifixion. There can be no doubt about Strauss's being right as to the Christian intensity of conviction, strenuousness of assertion, and readiness to suffer for the sake of their faith in Christ; and these are the main points with which we are concerned. We arrive therefore at the conclusion that the first Christians were sufficiently unanimous, coherent and undaunted to convince the foremost of their enemies. They were not so before the Crucifixion; they could not certainly have been made so by the Crucifixion alone; something beyond the Crucifixion must have occurred to give them such a moral ascendancy as should suffice to generate a revulsion of feeling in the mind of the persecuting Saul. Strauss asks us to believe that this missing something is to be found in the hallucinations of two or three men whose names have not been recorded and who have left no mark of their own. Is there any occasion for answer?

It is inconceivable that he who could write the Epistle to the Romans should not also have been as able as any man who ever lived to question the early believers as to their converse with Christ, and to report faithfully the substance of what they told him. That he knew the other Apostles, that he went up to Jerusalem to hold conferences with them, that he abode fifteen days with St. Peter—as he tells us, in order "to question him"—these things are certain. The Greek word ιστορησαι is a very suggestive one. It is so easy to make too much out of anything that I hardly dare to say how strongly the use of the verb ιστορειν suggests to me "getting at the facts of the case," "questioning as to how things happened," yet such would be the most obvious meaning of the word from which our own "history" and "story" are derived. Fifteen days was time enough to give Paul the means of coming to an understanding with Peter as to what the value of Peter's story was, nor can we believe that Paul should not both receive and transmit perfectly all that he was then told. In fact, without supposing these men to be so utterly visionary that nothing durable could come out of them, there is no escape from holding that Peter was justified in firmly believing that he had seen Christ alive within a very few days of the Crucifixion, that he succeeded also in satisfying Paul that this belief was well-founded, and that in the account of Christ's reappearances, as given I. Cor. xv., we have a virtually verbatim report of what Paul heard from Peter and the other Apostles. Of course the possibility remains that Paul may have been too easily satisfied, and not have cross-examined Peter as closely as he might have done. But then Paul was converted before this interview; and this implies that he had already found a general consent among the Christians whom he had met with, that the story which he afterwards heard from Peter (or one to the same effect) was true. Whence then the unanimity of this belief? Strauss answers as before—from the hallucinations of an originally small minority. We can only again reply that for the reasons already given we find it quite impossible to agree with him.

[The quotation from Strauss given in this chapter will be found pp. 414, 415, 420, of the first volume of the English translation, published by Williams and Norgate, 1865. I believe that my brother intended to make a fresh translation from the original passages, but he never carried out his intention, and in his MS. the page of the English translation with the first and last words of each passage are alone given. I could hardly venture to undertake the responsibility of making a fresh translation myself, and have therefore adhered almost word for word to the published English translation—here and there, however, a trifling alteration was really irresistible on the scores alike of euphony and clearness.—W. B. O.]

CHAPTER IV

PAUL'S TESTIMONY CONSIDERED

Enough has perhaps been said to cause the reader to agree with the view of St. Paul's conversion taken above—that is to say, to make him regard the conversion as mainly, if not entirely, due to the weight of evidence afforded by the courage and consistency of the early Christians.

But, the change in Paul's mind being thus referred to causes which preclude all possibility of hallucination or ecstasy on his own part, it becomes unnecessary to discuss the attempts which have been made to explain away the miraculous character of the account given in the Acts. I believe that this account is founded upon fact, and that it is derived from some description furnished by St. Paul himself of the vision mentioned, I. Cor. xv., which again is very possibly the same as that of II. Cor. xii. For the purposes of the present investigation, however, the whole story must be set aside. At the same time it should be borne in mind, that any detraction from the historical accuracy of the writer of the Acts, is more than compensated for, by the additional weight given to the conversion of St. Paul, whom we are now able to regard as having been converted by evidence which was in itself overpowering, and which did not stand in need of any miraculous interference in order to confirm it.

It is important to observe that the testimony of Paul should carry more weight with those who are bent upon close critical investigation than that even of the Evangelists. St. Paul is one whom we know, and know well. No syllable of suspicion has ever been breathed, even in Germany, against the first four of the Epistles which have been generally assigned to him; friends and foes of Christianity are alike agreed to accept them as the genuine work of the Apostle. Few figures, therefore, in ancient history stand out more clearly revealed to us than that of St. Paul, whereas a thick veil of darkness hangs over that of each one of the Evangelists. Who St. Matthew was, and whether the gospel that we have is an original work, or a translation (as would appear from Papias, our highest authority), and how far it has been modified in translation, are things which we shall never know. The Gospels of St. Mark and St. Luke are involved in even greater obscurity. The authorship, date, and origin of the fourth Gospel have been, and are being, even more hotly contested than those of the other three, and all that can be affirmed with certainty concerning it is, that no trace of its existence can be found before the latter half of the second century, and that the spirit of the work itself is eminently anti-Judaistic, whereas St. John appears both from the Gospels and from St. Paul's Epistles to have been a pillar of Judaism.

With St. Paul all is changed: we not only know him better than we know nine-tenths of our own most eminent countrymen of the last century, but we feel a confidence in him which grows greater and greater the more we study his character. He combines to perfection the qualities that make a good witness—capacity and integrity: add to this that his conclusions were forced upon him. We therefore feel that, whereas from a scientific point of view, the Gospel narratives can only be considered as the testimony of early and sincere writers of whom we know little or nothing, yet that in the evidence of St. Paul we find the missing link which connects us securely with actual eye-witnesses and gives us a confidence in the general accuracy of the Gospels which they could never of themselves alone have imparted. We could indeed ill spare either the testimony of the Evangelists or that of St. Paul, but if we were obliged to content ourselves with one only, we should choose the Apostle.

Turning then to the evidence of St. Paul as derivable from I. Cor. xv. we find the following:

"Moreover, brethren, I declare unto you the gospel which I preached unto you, which also ye have received and wherein ye stand. By which also ye are saved if ye keep in memory what I preached unto you, unless ye have believed in vain. For I delivered unto you first of all that which I also received, how that Christ died for our sins according to the Scriptures: and that He was buried, and that He rose again the third day according to the Scriptures; and that He was seen of Cephas, then of the twelve: after that He was seen of above five hundred brethren at once; of whom the greater portion remain unto this present, but some are fallen asleep. After that He was seen of James; then of all the Apostles. And last of all He was seen of me also, as of one born out of due time."

In the first place we must notice Paul's assertion that the Gospel which he was then writing was identical with that which he had originally preached. We may assume that each of the appearances of Christ here mentioned had in Paul's mind a definite time and place, derived from the account which he had received and which probably led to his conversion; the words "that which I also received" surely imply "that which I also received in the first instance": now we know from his own mouth (Gal. i., 16, 17) that after his conversion he "conferred not with flesh and blood"—"neither," he continues, "went I up to Jerusalem to them which were Apostles before me, but I went into Arabia, and returned again unto Damascus: then after three years I went up to Jerusalem to see (ιστορησαι) Peter, and abode with him fifteen days, but others of the Apostles saw I none, save James the Lord's brother." Since, then, he must have heard some story concerning Christ's reappearances before his conversion and subsequent sojourn in Arabia, and since he had heard nothing from eye-witnesses until the time of his going up to Jerusalem three years later, it is probable that the account quoted above is the substance of what he found persisted in by the Christians whom he was persecuting at Damascus, and was at length compelled to believe. But this is very unimportant: it is more to the point to insist upon the fact that St. Paul must have received the account given I. Cor. xv., 3–8 within a very few years of the Crucifixion itself, and that it was subsequently confirmed to him by Peter, and probably by James and John, during his stay of fifteen days in Peter's house.

This account can have been nothing new even then, for it is plain that at the time of Paul's conversion the Christian Church had spread far: Paul speaks of returning to Damascus, as though the writer of the Acts was right as regards the place of his conversion; but the fact of there having been a church in Damascus of sufficient importance for Paul to go thither to persecute it, involves the lapse of considerable time since the original promulgation of our Lord's Resurrection, and throws back the origin of the belief in that event to a time closely consequent upon the Crucifixion itself.

Now Paul informs us that he was told (we may assume by Peter and James) that Christ first reappeared within three days of the Crucifixion. There is no sufficient reason for doubting this; and one fact of weekly recurrence even to this day, affords it striking confirmation—I refer to the institution of Sunday as the Lord's day. We know that the observance of this day in commemoration of the Resurrection was a very early practice, nor is there anything which would seem to throw doubt upon the fact of the first "Sunday" having been also the Sunday of the Resurrection. Another confirmation of the early date assigned to the Resurrection by St. Paul, is to be found in the fact that every instinct would warn the Apostles against the third day as being dangerously early, and as opening a door for the denial of the completeness of the death. The fortieth day would far more naturally have been chosen.

Turning now from the question of the date of the first reappearance to what is told us of the reappearances themselves, we find that the earliest was vouchsafed to St. Peter, which is at first sight opposed to the Evangelistic records; but this is a discrepancy upon which no stress should be laid; St. Paul might well be aware that Mary Magdalene was the first to look upon her risen Lord, and yet have preferred to dwell upon the more widely known names of Peter and his fellow Apostles. The facts are probably these, that our Lord first shewed Himself to the women, but that Peter was the first of the Apostolic body to see Him; it was natural that if our Lord did not choose to show Himself to the Apostles without preparation, Peter should have been chosen as the one best fitted to prepare them: Peter probably collected the other Apostles, and then the Redeemer shewed Himself alive to all together. This is what we should gather from St. Paul's narrative; a narrative which it would seem arbitrary to set aside in the face of St. Paul's character, opportunities and antecedent prejudices against Christianity—in the face also of the unanimity of all the records we have, as well as of the fact that the Christian religion triumphed, and of the endless difficulties attendant on the hallucination theory.

We conclude therefore that Paul was satisfied by sufficient evidence that our Lord had appeared to Peter on the third day after the Crucifixion, nor can any reasonable doubt be thrown upon the other appearances of which he tells us. It is true that on the occasion of his visit to Peter he saw none other of the Apostles save James—but there is nothing to lead us to suppose that there was any want of unanimity among them: no trace of this has come down to us, and would surely have done so if it had existed. If any dependence at all is to be placed on the writers of the New Testament it did not exist. Stronger evidence than this unanimity it would be hard to find.

Another most noticeable feature is the fewness of the recorded appearances of Christ. They commenced according to Paul (and this is virtually according to Peter and James) immediately after the Crucifixion. Paul mentions only five appearances: this does not preclude the supposition that he knew of more, nor that the women who came to the sepulchre had also seen Him, but it does seem to imply that the reappearances were few in number, and that they continued only for a very short time. They were sufficient for their purpose: one of preparation to Peter—another to the Apostles—another to the outside world, and then one or two more—but still not more than enough to establish the fact beyond all possibility of dispute. The writer of the Acts tells us that Christ was seen for a space of forty days— presumably not every day, but from time to time. Now forty days is a mystical period, and one which may mean either more or less, within a week or two, than the precise time stated; it seems upon the whole most reasonable to conclude that the reappearances recorded by Paul, and some few others not recorded, extended over a period of one or two months after the Crucifixion, and that they then came to an end; for there can be no doubt that St. Paul conceived them as having ended with the appearance to the assembled Apostles mentioned I. Cor. xv., 7, and, though he does not say so expressly, there is that in the context which suggests their having been confined to a short space of time.

It is perfectly clear that St. Paul did not believe that any one had seen Christ in the interval between the last recorded appearance to the eleven, and the vision granted to himself. The words "and last of all he was seen also of me as of one born out of due time" point strongly in the direction of a lapse of some years between the second appearance to the eleven and his own vision. This confirms and is confirmed by the writer of the Acts. St. Paul never could have used the words quoted above, if he had held that the appearances which he records had been spread over a space of years intervening between the Crucifixion and his own vision. Where would be the force of "born out of due time" unless the time of the previous appearances had long passed by? But if, at the time of St. Paul's conversion, it was already many years since the last occasion upon which Christ had been seen by his disciples, we find ourselves driven back to a time closely consequent upon the Crucifixion as the only possible date of the reappearances. But this is in itself sufficient condemnation of Strauss's theory: that theory requires considerable time for the development of a perfectly unanimous and harmonious belief in the hallucinations, while every particle of evidence which we can get points in the direction of the belief in the Resurrection having followed very closely upon the Crucifixion.

To repeat: had the reappearances been due to hallucination only, they would neither have been so few in number nor have come to an end so soon. When once the mind has begun to run riot in hallucination, it is prodigal of its own inventions. Favoured believers would have been constantly seeing Christ even up to the time of Paul's letter to the Corinthians, and the Apostle would have written that even then Christ was still occasionally seen of those who trusted in him, and served him faithfully. But we meet with nothing of the sort: we are told that Christ was seen a few times shortly after the Crucifixion, then after a lapse of several years (I am surely warranted in saying this) Paul himself saw Him—but no one in the interval, and no one afterwards. This is not the manner of the hallucinations of uneducated people. It is altogether too sober: the state of mind from which alone so baseless a delusion could spring, is one which never could have been contented with the results which were evidently all, or nearly all, that Paul knew of. St. Paul's words cannot be set aside without more cause than Strauss has shewn: instead of betraying a tendency towards exaggeration, they contain nothing whatever, with the exception of his own vision, that is not imperatively demanded in order to account for the rise and spread of Christianity.

Concerning that vision Strauss writes as follows:

"With regard to the appearance he (Paul) witnessed—he uses the same word (ὤφθη) as with regard to the others: he places it in the same category with them only in the last place, as he names himself the last of the Apostles, but in exactly the same rank with the others. Thus much, therefore, Paul knew—or supposed—that the appearances which the elder disciples had seen soon after the Resurrection of Jesus had been of the same kind as that which had been, only later, vouchsafed to himself. Of what sort then was this?"

I confess that I am wholly unable to feel the force of the above. Strauss says that Paul's vision was ecstatic—subjective and not objective—that Paul thought he saw Christ, although he never really saw him. But, says Strauss, he uses the same word for his own vision and for the appearances to the earlier Apostles: it is plain therefore that he did not suppose the earlier Apostles to have seen Christ in the same sort of way in which they saw themselves and other people, but to have seen him as Paul himself did, i.e., by supernatural revelation.

But would it not be more fair to say that Paul's using the same word for all the appearances—his own vision included—implies that he considered this last to have been no less real than those vouchsafed earlier, though he may have been perfectly well aware that it was different in kind? The use of the same word for all the appearances is quite compatible with a belief in Paul's mind that the manner in which he saw Christ was different from that in which the Apostles had seen him: indeed, so long as he believed that he had seen Christ no less really than the others, one cannot see why he should have used any other word for his own vision than that which he had applied to the others: we should even expect that he would do so, and should be surprised at his having done otherwise. That Paul did believe in the reality of his own vision is indisputable, and his use of the word ὤφθη was probably dictated by a desire to assert this belief in the strongest possible way, and to place his own vision in the same category with others, which were so universally known among Christians to have been material and objective, that there was no occasion to say so. Nevertheless there is that in Paul's words on which Strauss does not dwell, but which cannot be passed over without notice. Paul does not simply say, "and last of all he was seen also of me"—but he adds the words "as of one born out of due time."

It is impossible to say decisively that this addition implies that Paul recognised a difference in kind between the appearances, inasmuch as the words added may only refer to time—still they would explain the possible use of [ωφθη] in a somewhat different sense, and I cannot but think that they will suggest this possibility to the reader. They will make him feel, if he does not feel it without them, how strained a proceeding it is to bind Paul down to a rigorously identical meaning on every occasion on which the same word came from his pen, and to maintain that because he once uses it on the occasion of an appearance which he held to be vouchsafed by revelation, therefore, wherever else he uses it, he must have intended to refer to something seen by revelation: the words "as of one born out of due time" imply the utterly unlooked for and transcendent nature of the favour, and suggest, even though they do not compel, the inference that while the other Apostles had seen Christ in the common course of nature, as a visible tangible being before their waking eyes, he had himself seen Him not less truly, but still only by special and unlooked for revelation. If such thoughts were in his mind he would not probably have expressed them farther than by the touching words which he has added concerning his own vision. So much for the objection that the evidence of Paul concerning the earlier appearances is impaired by his having used the same word for them, and for the appearance to himself. It only remains therefore to review in brief the general bearings of Paul's testimony as given I. Cor. xv., 1–8.

Firstly, there is the early commencement of the reappearances: this is incompatible with hallucination, for the hallucination must be supposed to have occurred when most easy to refute, and when the spell of shame and fear was laid most heavily upon the Apostles. Strauss maintains that the appearances were unconsciously antedated by Peter; we can only say that the circumstances of the case, as entered into more fully above, render this very improbable; that if Peter told Paul that he saw Christ on the third day after the Crucifixion, he probably firmly believed that he did see Him; and that if he believed this, he was also probably right in so believing.

Secondly, there is the fact that the reappearances were few, and extended over a short time only. Had they been due to hallucination there would have been no limit either to their number or duration. Paul seems to have had no idea that there ever had been, or ever would be, successors to the five hundred brethren who saw Christ at one time. Some were fallen asleep—the rest would in time follow them. It is incredible that men should have so lost all count of fact, so debauched their perception of external objects, so steeped themselves in belief in dreams which had no foundation but in their own disordered brains, as to have turned the whole world after them by the sheer force of their conviction of the truth of their delusions, and yet that suddenly, within a few weeks from the commencement of this intoxication, they should have come to a dead stop and given no further sign of like extravagance. The hallucinations must have been so baseless, and would argue such an utter subordination of judgement to imagination, that instead of ceasing they must infallibly have ended in riot and disorganisation; the fact that they did cease (which cannot be denied) and that they were followed by no disorder, but by a solemn sober steadfastness of purpose, as of reasonable men in deadly earnest about a matter which had come to their knowledge, and which they held it vital for all to know—this fact alone would be sufficient to overthrow the hallucination theory. Such intemperance could never have begotten such temperance: from such a frame of mind as Strauss assigns to the Apostles no religion could have come which should satisfy the highest spiritual needs of the most civilised nations of the earth for nearly two thousand years.

When, therefore, we look at the want of faith of the Apostles before the Crucifixion, and to their subsequent intense devotion; at their unanimity at their general sobriety; at the fact that they succeeded in convincing the ablest of their enemies and ultimately the whole of Europe; at the undeviating consent of all the records we have; at the early date at which the reappearances commenced,—at their small number and short duration—things so foreign to the nature of hallucination; at the excellent opportunities which Paul had for knowing what he tells us; at the plain manner in which he tells it, and the more than proof which he gave of his own conviction of its truth; at the impossibility of accounting for the rise of Christianity without the reappearance of its Founder after His Crucifixion; when we look at all these things we shall admit that it is impossible to avoid the belief that after having died, Christ did reappear to his disciples, and that in this fact we have the only intelligible explanation of the triumph of Christianity.

CHAPTER V

A CONSIDERATION OF CERTAIN ILL-JUDGED METHODS OF DEFENCE

The reader has now heard the utmost that can be said against the historic character of the Resurrection by the ablest of its impugners. I know of nothing in any of Strauss's works which can be considered as doing better justice to his opinions than the passages which I have quoted and, I trust, refuted. I have quoted fully, and have kept nothing in the background. If I had known of anything stronger against the Resurrection from any other source, I should certainly have produced it. I have answered in outline only, but I do not believe that I have passed any difficulty on one side.

What then does the reader think? Was the attack so dangerous, or the defence so far to seek? I believe he will agree with me that the combat was one of no great danger when it was once fairly entered upon. But the wonder, and, let me add, the disgrace, to English divines, is that the battle should have been shirked so long. What is it that has made the name of Strauss so terrible to the ears of English Churchmen? Surely nothing but the ominous silence which has been maintained concerning him in almost all quarters of our Church. For what can he say or do against the other miracles if he be powerless against the Resurrection? He can make sentences which sound plausible, but that is no great feat. Can he show that there is any a priori improbability whatever, in the fact of miracles having been wrought by one who died and rose from the dead? If a man did this it is a small thing that he should also walk upon the waves and command the winds. But if there is no a priori difficulty with regard to these miracles, there is certainly none other.

Let this, however, for the present pass, only let me beg of the reader to have patience while I follow out the plan which I have pursued up to the present point, and proceed to examine certain difficulties of another character. I propose to do so with the same unflinching examination as heretofore, concealing nothing that has been said, or that can be said; going out of my way to find arguments for opponents, if I do not think that they have put forward all that from their own point of view they might have done, and careless how many difficulties I may bring before the reader which may never yet have occurred to him, provided I feel that I can also shew him how little occasion there is to fear them.

I must, however, maintain two propositions, which may perhaps be unfamiliar to some of those who have not as yet given more than a conventional and superficial attention to the Scriptural records, but which will meet with ready assent from all whose studies have been deeper. Fain would I avoid paining even a single reader, but I am convinced that the arresting of infidelity depends mainly upon the general recognition of two broad facts. The first is this—that the Apostles, even after they had received the gift of the Holy Spirit were still fallible though holy men; the second—that there are certain passages in each of the Gospels as we now have them, which were not originally to be found therein, and others which, though genuine, are still not historic. This much of concession we must be prepared to make, and we shall find (as in the case of the conversion of St. Paul) that our position is indefinitely strengthened by doing so.

When shall we Christians learn that the truest ground is also the strongest? We may be sure that until we have done so we shall find a host of enemies who will say that truth is not ours. It is we who have created infidelity, and who are responsible for it. We are the true infidels, for we have not sufficient faith in our own creed to believe that it will bear the removal of the incrustations of time and superstition. When men see our cowardice, what can they think but that we must know that we have cause to be afraid? We drive men into unbelief in spite of themselves, by our tenacious adherence to opinions which every unprejudiced person must see at a glance that we cannot rightfully defend, and then we pride ourselves upon our love for Christ and our hatred of His enemies. If Christ accepts this kind of love He is not such as He has declared Himself.

We mistake our love of our own immediate ease for the love of Christ, and our hatred of every opinion which is strange to us, for zeal against His enemies. If those to whom the unfamiliarity of an opinion or its inconvenience to themselves is a test of its hatefulness to Christ, had been born Jews, they would have crucified Him whom they imagine that they are now serving: if Turks, they would have massacred both Jew and Christian; if Papists at the time of the Reformation they would have persecuted Protestants: if Protestants, under Elizabeth, Papists. Truth is to them an accident of birth and training, and the Christian faith is in their eyes true because these accidents, as far as they are concerned, have decided in its favour. But such persons are not Christians. It is they who crucify Christ, who drive men from coming to Him whose every instinct would lead them to love and worship Him, but who are warned off by observing the crowd of sycophants and time-servers who presume to call Him Lord.

But to look at the matter from another point of view; when there is a long sustained contest between two bodies of capable and seriously disposed people, (and none can deny that many of our adversaries have been both one and the other), and when this contest shews no sign of healing, but rather widens from generation to generation, and each party accuses the other of disingenuousness, obstinacy and other like serious defects of mind—it may be certainly assumed that the truth lies wholly with neither side, but that each should make some concessions to the other. A third party sees this at a glance, and is amazed because neither of the disputants can perceive that his opponent must be possessed of some truths, in spite of his trying to defend other positions which are indefensible. Strange! that a thing which it seems so easy to avoid, should so seldom be avoided! Homer said well:

"Perish strife, both from among gods and men, And wrath which maketh even him that is considerate, cruel, Which getteth up in the heart of a man like smoke, And the taste thereof is sweeter than drops of honey."

But strife can never cease without concessions upon both sides. We agree to this readily in the abstract, but we seldom do so when any given concession is in question. We are all for concession in the general, but for none in the particular, as people who say that they will retrench when they are living beyond their income, but will not consent to any proposed retrenchment. Thus many shake their heads and say that it is impossible to live in the present age and not be aware of many difficulties in connection with the Christian religion; they have studied the question more deeply than perhaps the unbeliever imagines; and having said this much they give themselves credit for being wide-minded, liberal and above vulgar prejudices: but when pressed as to this or that particular difficulty, and asked to own that such and such an objection of the infidel's needs explanation, they will have none of it, and will in nine cases out of ten betray by their answers that they neither know nor want to know what the infidel means, but on the contrary that they are resolute to remain in ignorance. I know this kind of liberality exceedingly well, and have ever found it to harbour more selfishness, idleness, cowardice and stupidity than does open bigotry. The bigot is generally better than his expressed opinions, these people are invariably worse than theirs.

The above principle has been largely applied in the writings of so-called orthodox commentators, not unfrequently even by men who might have been assumed to be above condescending to such trickery. A great preface concerning candour, with a flourish of trumpets in the praise of truth, seems to have exhausted every atom of truth and candour from the work that follows it.

It will be said that I ought not to make use of language such as this without bringing forward examples. I shall therefore adduce them.

One of the most serious difficulties to the unbeliever is the inextricable confusion in which the accounts of the Resurrection have reached us: no one can reconcile these accounts with one another, not only in minute particulars, but in matters on which it is of the highest importance to come to a clear understanding. Thus, to omit all notice of many other discrepancies, the accounts of Mark, Luke, and John concur in stating that when the women came to the tomb of Jesus very early on the Sunday morning, they found it already empty: the stone was gone when they came there, and, according to John, there was not even an angelic vision for some time afterwards. There is nothing in any of these three accounts to preclude the possibility of the stone's having been removed within an hour or two of the body's having been laid in the tomb.

But when we turn to Matthew we find all changed: we are told that the stone was gone not when the women came, but that on their arrival there was a great earthquake, and that an angel came down from Heaven, and rolled away the stone, and sat upon it, and that the guard who had been set over the tomb (of whom we hear nothing from any of the other evangelists) became as dead men while the angel addressed the women.

Now this is not one of those cases in which the supposition can be tolerated that all would be clear if the whole facts of the case were known to us. No additional facts can make it come about that the tomb should have been sealed and guarded, and yet not sealed and guarded; that the same women, at the same time and place, should have witnessed an earthquake, and yet not witnessed one; have found a stone already gone from a tomb, and yet not found it gone; have seen it rolled away, and not seen it, and so on; those who say that we should find no difficulty if we knew all the facts are still careful to abstain from any example (so far as I know) of the sort of additional facts which would serve their purpose. They cannot give one; any mind which is truly candid—white—not scrawled and scribbled over till no character is decipherable—will feel at once that the only question to be raised is, which is the more correct account of the Resurrection—Matthew's or those given by the other three Evangelists? How far is Matthew's account true, and how far is it exaggerated? For there must be either exaggeration or invention somewhere. It is inconceivable that the other writers should have known the story told by Matthew, and yet not only made no allusion to it, but introduced matter which flatly contradicts it, and it is also inconceivable that the story should be true, and yet that the other writers should not have known it.

This is how the difficulty stands—a difficulty which vanishes in a moment if it be rightly dealt with, but which, when treated after our unskilful English method, becomes capable of doing inconceivable mischief to the Christian religion. Let us see then what Dean Alford—a writer whose professions of candour and talk about the duty of unflinching examination leave nothing to be desired—has to say upon this point. I will first quote the passage in full from Matthew, and then give the Dean's note. I have drawn the greater part of the comments that will follow it from an anonymous pamphlet {141} upon the Resurrection, dated 1865, but without a publisher's name, so that I presume it must have been printed for private circulation only.

St. Matthew's account runs:—

"Now the next day, that followed the day of the preparation, the chief priests and Pharisees came together unto Pilate, saying, 'Sir, we remember that that deceiver said, while he was yet alive, "After three days I will rise again." Command therefore that the sepulchre be made sure until the third day, lest his disciples come by night and steal him away and say unto the people, "He is risen from the dead:" so the last error shall be worse than the first.' Pilate said unto them, 'Ye have a watch: go your way, make it as sure as ye can.' So they went and made the sepulchre sure, sealing the stone and setting a watch. In the end of the Sabbath, as it began to dawn towards the first day of the week, came Mary Magdalene and the other Mary to see the sepulchre. And, behold, there was a great earthquake: for the angel of the Lord descended from heaven, and came and rolled back the stone from the door, and sat upon it. His countenance was like lightning, and his raiment white as snow: And for fear of him the keepers did shake, and became as dead men. And the angel answered and said unto the women, 'Fear not ye: for I know that ye seek Jesus, which was crucified. He is not here: for he is risen, as he said. Come, see the place where the Lord lay. And go quickly, and tell his disciples that he is risen from the dead; and, behold, he goeth before you into Galilee; there shall ye see him: lo, I have told you.' And they departed quickly from the sepulchre with fear and great joy; and did run to bring his disciples word. And as they went to tell his disciples, Jesus met them, saying, 'All hail.' And they came and held him by the feet, and worshipped him (cf. John xx., 16, 17). Then said Jesus unto them, 'Be not afraid: go tell my brethren that they go into Galilee, and there shall they see me.' Now when they were going, behold, some of the watch came into the city, and shewed unto the chief priests all the things that were done. And when they were assembled with the elders, and had taken counsel, they gave large money unto the soldiers, saying, 'Say ye, His disciples came by night, and stole him away while we slept. And if this come to the governor's ears, we will persuade him and secure you.' So they took the money, and did as they were taught: and this saying is commonly reported among the Jews until this day."

Let us turn now to the Dean's note on Matt. xxvii., 62–66.

With regard to the setting of the watch and sealing of the stone, he tells us that the narrative following (i.e., the account of the guard and the earthquake) "has been much impugned and its historical accuracy very generally given up even by the best of the German commentators (Olshausen, Meyer; also De Wette, Hase, and others). The chief difficulties found in it seem to be: (1) How should the chief priests, &c., know of His having said 'in three days I will rise again,' when the saying was hid even from His own disciples? The answer to this is easy. The meaning of the saying may have been, and was hid from the disciples; but the fact of its having been said could be no secret. Not to lay any stress on John ii., 19 (Jesus answered and said unto them, 'Destroy this temple and in three days I will build it up'), we have the direct prophecy of Matt. xii., 40 ('For as Jonah was three days and three nights in the whale's belly, so shall the Son of Man be three days and three nights in the heart of the earth): besides this there would be a rumour current, through the intercourse of the Apostles with others, that He had been in the habit of so saying. (From what source can Dean Alford know that our Lord was in the habit of so saying? What particle of authority is there for this alleged habit of our Lord?) As to the understanding of the words we must remember that hatred is keener sighted than love: that the raising of Lazarus would shew what sort of a thing rising from the dead was to be; and the fulfilment of the Lord's announcement of his crucifixion would naturally lead them to look further to what more he had announced. (2) How should the women who were solicitous about the removal of the stone not have been still more so about its being sealed and a guard set? The answer to this last has been given above—they were not aware of the circumstance because the guard was not set till the evening before. There would be no need of the application before the approach of the third day—it is only made for a watch, εως της τρίτης ημέρας (ver. 64), and it is not probable that the circumstance would transpire that night—certainly it seems not to have done so. (3) That Gamaliel was of the council, and if such a thing as this and its sequel (chap. xxviii., 11–15) had really happened, he need not have expressed himself doubtfully (Acts v., 39), but would have been certain that this was from God. But, first, it does not necessarily follow that every member of the Sanhedrim was present, and applied to Pilate, or even had they done so, that all bore a part in the act of xxviii., 12" (the bribing of the guard to silence). "One who like Joseph had not consented to the deed before—and we may safely say that there were others such—would naturally withdraw himself from further proceedings against the person of Jesus. (4) Had this been so the three other Evangelists would not have passed over so important a testimony to the Resurrection. But surely we cannot argue in this way—for thus every important fact narrated by one Evangelist alone must be rejected, e.g. (which stands in much the same relation), the satisfaction of Thomas—another such narrations. Till we know more about the circumstances under which, and the scope with which, each Gospel was compiled, all a priori arguments of this kind are good for nothing."

(The italics in the above, and throughout the notes quoted, are the Dean's, unless it is expressly stated otherwise.)

I will now proceed to consider this defence of Matthew's accuracy against the objections of the German commentators.

I. The German commentators maintain that the chief priests are not likely to have known of any prophecy of Christ's Resurrection when His own disciples had evidently heard of nothing to this effect. Dean Alford's answer amounts to this:—

1. They had heard the words but did not understand their meaning; hatred enabled the chief priests to see clearly what love did not reveal to the understanding of the Apostles. True, according to Matthew, Christ had said that as Jonah was three days and three nights in the whale's belly, so the Son of Man should be three days and three nights in the heart of the earth; but it would be only hatred which would suggest the interpretation of so obscure a prophecy: love would not be sufficiently keen-sighted to understand it.

But in the first place I would urge that if the Apostles had ever heard any words capable of suggesting the idea that Christ should rise, after they had already seen the raising of Lazarus, on whom corruption had begun its work, they must have expected the Resurrection. After having seen so stupendous a miracle, any one would expect anything which was even suggested by the One who had performed it. And, secondly, hatred is not keener sighted than love.

2. Dean Alford says that the raising of Lazarus would shew the chief priests what sort of a thing the Resurrection from the dead was to be, and that the fulfilment of Christ's prophecy concerning his Crucifixion would naturally lead them to look further to what else he had announced.

But, if the raising of Lazarus would shew the chief priests what sort of thing the Resurrection was to be, it would shew the Apostles also; and again if the fulfilment of the prophecy of the Crucifixion would lead the chief priests to look further to the fulfilment of the prophecy of the Resurrection, so would it lead the Apostles; this supposition of one set of men who can see everything, and of another with precisely the same opportunities and no less interest, who can see nothing, is vastly convenient upon the stage, but it is not supported by a reference to Nature; self-interest would have opened the eyes of the Apostles.

II. The German commentators ask how was it possible that the women who were solicitous about the removal of the stone, should not be still more so about "its being sealed and a guard set?" If the German commentators have asked their question in this shape, they have asked it badly, and Dean Alford's answer is sufficient: they might have asked, how the other three writers could all tell us that the stone was already gone when the women got there, and yet Matthew's story be true? and how Matthew's story could be true without the other writers having known it? and how the other writers could have introduced matter contradictory to it, if they had known it to be true?

III. The German commentators say that in the Acts of the Apostles we find Gamaliel expressing himself as doubtful whether or no Christianity was of God, whereas had he known the facts related by Matthew he could have had no doubt at all. He must have known that Christianity was of God.

Dean Alford answers that perhaps Gamaliel was not there. To which I would rejoin that though Gamaliel might have had no hand in the bribery, supposing it to have taken place, it is inconceivable that such a story should have not reached him; the matter could never have been kept so quiet but that it must have leaked out. Men are not so utterly bad or so utterly foolish as Dean Alford seems to imply; and whether Gamaliel was or was not present when the guard were bribed, he must have been equally aware of the fact before making the speech which is assigned to him in the Acts.

IV. The German commentators argue from the silence of the other Evangelists: Dean Alford replies by denying that this silence is any argument: but I would answer, that on a matter which the other three writers must have known to have been of such intense interest, their silence is a conclusive proof either of their ignorance or their indolence as historians. Dean Alford has well substantiated the independence of the four narratives, he has well proved that the writer of the fourth Gospel could never have seen the other Gospels, and yet he supposes that that writer either did not know the facts related by Matthew, or thought it unnecessary to allude to them. Neither of these suppositions is tenable: but there would nevertheless be a shadow of ground for Dean Alford to stand upon if the other Evangelists were simply silent: but why does he omit all notice of their introducing matter which is absolutely incompatible with Matthew's accuracy?

There is one other consideration which must suggest itself to the reader in connection with this story of the guard. It refers to the conduct of the chief priests and the soldiers themselves. The conduct assigned to the chief priests in bribing the guard to lie against one whom they must by this time have known to be under supernatural protection, is contrary to human nature. The chief priests (according to Matthew) knew that Christ had said he should rise: in spite of their being well aware that Christ had raised Lazarus from the dead but very recently they did not believe that he would rise, but feared (so Matthew says) that the Apostles would steal the body and pretend a resurrection: up to this point we admit that the story, though very improbable, is still possible: but when we read of their bribing the guards to tell a lie under such circumstances as those which we are told had just occurred, we say that such conduct is impossible: men are too great cowards to be capable of it. The same applies to the soldiers: they would never dare to run counter to an agency which had nearly killed them with fright on that very selfsame morning. Let any man put himself in their position: let him remember that these soldiers were previously no enemies to Christ, nor, as far as we can judge, is it likely that they were a gang of double-dyed villains: but even if they were, they would not have dared to act as Matthew says they acted.

And now let us turn to another note of Dean Alford's.

Speaking of the independence of the four narratives (in his note on Matt. xxviii., 1–10) and referring to their "minor discrepancies," the Dean says, "Supposing us to be acquainted with every thing said and done in its order and exactness, we should doubtless be able to reconcile, or account for, the present forms of the narratives; but not having this key to the harmonising of them, all attempts to do so in minute particulars must be full of arbitrary assumptions, and carry no certainty with them: and I may remark that of all harmonies those of the incidents of these chapters are to me the most unsatisfactory. Giving their compilers all credit for the best intentions, I confess they seem to me to weaken instead of strengthening the evidence, which now rests (speaking merely objectively) on the unexceptionable testimony of three independent narrators, and one who besides was an eye witness of much that happened. If we are to compare the four and ask which is to be taken as most nearly reporting the exact words and incidents, on this there can, I think, be no doubt. On internal as well as external ground that of John takes the highest place, but not of course to the exclusion of those parts of the narrative which he does not touch."

Surely the above is a very extraordinary note. The difficulty of the irreconcilable differences between the four narratives is not met nor attempted to be met: the Dean seems to consider the attempt as hopeless: no one, according to him, has been as yet successful, neither can he see any prospect of succeeding better himself: the expedient therefore which he proposes is that the whole should be taken on trust; that it should be assumed that no discrepancy which could not be accounted for would be found, if the facts were known in the exact order in which they occurred. In other words, he leaves the difficulty where it was. Yet surely it is a very grave one. The same events are recorded by three writers (one being professedly an eye-witness, and the others independent writers), in a way which is virtually the same, in spite of some unimportant variations in the manner of telling it, while a fourth gives a totally different and irreconcilable account; the matter stands in such confusion at present that even Dean Alford admits that any attempt to reconcile the differences leaves them in worse confusion than ever; the ablest and most spiritually minded of the German commentators suggest a way of escape; nevertheless, according to the Dean we are not to profit by it, but shall avoid the difficulty better by a simpler process—the process of passing it over.

A man does well to be angry when he sees so solemn and momentous a subject treated thus. What is trifling if this is not trifling? What is disingenuousness if not this? It involves some trouble and apparent danger to admit that the same thing has happened to the Christian records which has happened to all others—i.e., that they have suffered—miraculously little, but still something—at the hands of time; people would have to familiarise themselves with new ideas, and this can seldom be done without a certain amount of wrangling, disturbance, and unsettling of comfortable ease: it is therefore by all means and at all risks to be avoided. Who can doubt that some such feeling as this was in Dean Alford's mind when the notes above criticised were written? Yet what are the means taken to avoid the recognition of obvious truth? They are disingenuous in the very highest degree. Can this prosper? Not if Christ is true.

What is the practical result? The loss of many souls who would gladly come to the Saviour, but who are frightened off by seeing the manner in which his case is defended. And what after all is the danger that would follow upon candour? None. Not one particle. Nevertheless, danger or no danger, we are bound to speak the truth. We have nothing to do with consequences and moral tendencies and risk to this or that fundamental principle of our belief, nor yet with the possibility of lurid lights being thrown here or there. What are these things to us? They are not our business or concern, but rest with the Being who has required of us that we should reverently, patiently, unostentatiously, yet resolutely, strive to find out what things are true and what false, and that we should give up all, rather than forsake our own convictions concerning the truth.

This is our plain and immediate duty, in pursuance of which we proceed to set aside the account of the Resurrection given in St. Matthew's Gospel. That account must be looked upon as the invention of some copyist, or possibly of the translator of the original work, at a time when men who had been eye-witnesses to the actual facts of the Resurrection were becoming scarce, and when it was felt that some more unmistakably miraculous account than that given in the other three Gospels would be a comfort and encouragement to succeeding generations. We, however, must now follow the example of "even the best" of the German commentators, and discard it as soon as possible. On having done this the whole difficulty of the confusion of the four accounts of the Resurrection vanishes like smoke, and we find ourselves with three independent writers whose differences are exactly those which we might expect, considering the time and circumstances in which they wrote, but which are still so trifling as to disturb no man's faith.

CHAPTER VI

MORE DISINGENUOUSNESS

[Here, perhaps, will be the fittest place for introducing a letter to my brother from a gentleman who is well known to the public, but who does not authorise me to give his name. I found this letter among my brother's papers, endorsed with the words "this must be attended to," but with nothing more. I imagine that my brother would have incorporated the substance of his correspondent's letter into this or the preceding chapter, but not venturing to do so myself, I have thought it best to give the letter and extract in full, and thus to let them speak for themselves.—W. B. O.]

June 15, 1868.

My dear Owen,

Your brother has told me what you are doing, and the general line of your argument. I am sorry that you should be doing it, for I need not tell you that I do not and cannot sympathise with the great and unexpected change in your opinions. You are the last man in the world from whom I should have expected such a change: but, as you well know, you are also the last man in the world whose sincerity in making it I should be inclined to question. May you find peace and happiness in whatever opinions you adopt, and let me trust also that you will never forget the lessons of toleration which you learnt as the disciple of what you will perhaps hardly pardon me for calling a freer and happier school of thought than the one to which you now believe yourself to belong.

Your brother tells me that you are ill; I need not say that I am sorry, and that I should not trouble you with any personal matter—I write solely in reference to the work which I hear that you have undertaken, and which I am given to understand consists mainly in the endeavour to conquer unbelief, by really entering into the difficulties felt by unbelievers. The scheme is a good one if thoroughly carried out. We imagine that we stand in no danger from any such course as this, and should heartily welcome any book which tried to grapple with us, even though it were to compel us to admit a great deal more than I at present think it likely that even you can extort from us. Much more should we welcome a work which made people understand us better than they do; this would indeed confer a lasting benefit both upon them and us.

However, I know you wish to do your work thoroughly; I want, therefore, to make a trifling suggestion which you will take pro tanto: it is this:—Paley, in his third book, professes to give "a brief consideration of some popular objections," and begins Chap. I. with "The discrepancies between the several Gospels."

Now, I know you have a Paley, but I know also that you are ill, and that people who are ill like being saved from small exertions. I have, therefore, bought a second-hand Paley for a shilling, and have cut out the chapter to which I especially want to call your attention. Will you kindly read it through from beginning to end?

Is it fair? Is the statement of our objections anything like what we should put forward ourselves? And can you believe that Paley with his profoundly critical instinct, and really great knowledge of the New Testament, should not have been perfectly well aware that he was misrepresenting and ignoring the objections which he professed to be removing?

He must have known very well that the principle of confirmation by discrepancy is one of very limited application, and that it will not cover anything approaching to such wide divergencies as those which are presented to us in the Gospels. Besides, how can he talk about Matthew's object as he does, and yet omit all allusion to the wide and important differences between his account of the Resurrection, and those of Mark, Luke, and John? Very few know what those differences really are, in spite of their having the Bible always open to them. I suppose that Paley felt pretty sure that his readers would be aware of no difficulty unless he chose to put them up to it, and wisely declined to do so. Very prudent, but very (as it seems to me) wicked. Now don't do this yourself. If you are going to meet us, meet us fairly, and let us have our say. Don't pretend to let us have our say while taking good care that we get no chance of saying it. I know you won't.

However, will you point out Paley's unfairness in heading this part of his work "A brief consideration of some popular objections," and then proceeding to give a chapter on "the discrepancies between the several Gospels," without going into the details of any of those important discrepancies which can have been known to none better than himself? This is the only place, so far as I remember, in his whole book, where he even touches upon the discrepancies in the Gospels. Does he do so as a man who felt that they were unimportant and could be approached with safety, or as one who is determined to carry the reader's attention away from them, and fix it upon something else by a coup de main?

This chapter alone has always convinced me that Paley did not believe in his own book. No one could have rested satisfied with it for moment, if he felt that he was on really strong ground. Besides, how insufficient for their purpose are his examples of discrepancies which do not impair the credibility of the main fact recorded!

How would it have been if Lord Clarendon and three other historians had each told us that the Marquis of Argyll came to life again after being beheaded, and then set to work to contradict each other hopelessly as to the manner of his reappearance? How if Burnet, Woodrow, and Heath had given an account which was not at all incompatible with a natural explanation of the whole matter, while Clarendon gave a circumstantial story in flat contradiction to all the others, and carefully excluded any but a supernatural explanation? Ought we to, or should we, allow the discrepancies to pass unchallenged? Not for an hour—if indeed we did not rather order the whole story out of court at once, as too wildly improbable to deserve a hearing.

You will, I know, see all this, and a great deal more, and will point it better than I can. Let me as an old friend entreat you not to pass this over, but to allow me to continue to think of you as I always have thought of you hitherto, namely, as the most impartial disputant in the world.—Yours, &c.

"I know not a more rash or unphilosophical conduct of the understanding, than to reject the substance of a story, by reason of some diversity in the circumstances with which it is related. The usual character of human testimony is substantial truth under circumstantial variety. This is what the daily experience of courts of justice teaches. When accounts of a transaction come from the mouths of different witnesses, it is seldom that it is not possible to pick out apparent or real inconsistencies between them. These inconsistencies are studiously displayed by an adverse pleader, but oftentimes with little impression upon the minds of the judges. On the contrary, close and minute agreement induces the suspicion of confederacy and fraud. When written histories touch upon the same scenes of action, the comparison almost always affords ground for a like reflection. Numerous and sometimes important variations present themselves; not seldom, also, absolute and final contradictions; yet neither one nor the other are deemed sufficient to shake the credibility of the main fact. The embassy of the Jews to deprecate the execution of Claudian's order to place his statue in their temple Philo places in harvest, Josephus in seed-time, both contemporary writers. No reader is led by this inconsistency to doubt whether such an embassy was sent, or whether such an order was given. Our own history supplies examples of the same kind. In the account of the Marquis of Argyll's death in the reign of Charles II., we have a very remarkable contradiction. Lord Clarendon relates that he was condemned to be hanged, which was performed the same day; on the contrary, Burnet, Woodrow, Heath, Echard, concur in stating that he was condemned upon the Saturday, and executed upon a Monday. {158a} Was any reader of English history ever sceptic enough to raise from hence a question, whether the Marquis of Argyll was executed or not? Yet this ought to be left in uncertainty, according to the principles upon which the Christian religion has sometimes been attacked. Dr. Middleton contended that the different hours of the day assigned to the Crucifixion of Christ by John and the other Evangelists, did not admit of the reconcilement which learned men had proposed; and then concludes the discussion with this hard remark: 'We must be forced, with several of the critics, to leave the difficulty just as we found it, chargeable with all the consequences of manifest inconsistency.' {158b} But what are these consequences? By no means the discrediting of the history as to the principal fact, by a repugnancy (even supposing that repugnancy not to be resolvable into different modes of computation) in the time of the day in which it is said to have taken place.

"A great deal of the discrepancy observable in the Gospels arises from omission; from a fact or a passage of Christ's life being noticed by one writer, which is unnoticed by another. Now, omission is at all times a very uncertain ground of objection. We perceive it not only in the comparison of different writers, but even in the same writer, when compared with himself. There are a great many particulars, and some of them of importance, mentioned by Josephus in his Antiquities, which as we should have supposed, ought to have been put down by him in their place in the Jewish Wars. {159a} Suetonius, Tacitus, Dion Cassius have all three written of the reign of Tiberius. Each has mentioned many things omitted by the rest, {159b} yet no objection is from thence taken to the respective credit of their histories. We have in our own times, if there were not something indecorous in the comparison, the life of an eminent person, written by three of his friends, in which there is very great variety in the incidents selected by them, some apparent, and perhaps some real, contradictions: yet without any impeachment of the substantial truth of their accounts, of the authenticity of the books, of the competent information or general fidelity of the writers.

"But these discrepancies will be still more numerous, when men do not write histories, but memoirs; which is perhaps the true name and proper description of our Gospels; that is, when they do not undertake, or ever meant to deliver, in order of time, a regular and complete account of all the things of importance which the person who is the subject of their history did or said; but only, out of many similar ones, to give such passages, or such actions and discourses, as offered themselves more immediately to their attention, came in the way of their enquiries, occurred to their recollection, or were suggested by their particular design at the time of writing.

"This particular design may appear sometimes, but not always, nor often. Thus I think that the particular design which St. Matthew had in view whilst he was writing the history of the Resurrection, was to attest the faithful performance of Christ's promise to his disciples to go before them into Galilee; because he alone, except Mark, who seems to have taken it from him, has recorded this promise, and he alone has confined his narrative to that single appearance to the disciples which fulfilled it. It was the preconcerted, the great and most public manifestation of our Lord's person. It was the thing which dwelt upon St. Matthew's mind, and he adapted his narrative to it. But, that there is nothing in St. Matthew's language which negatives other appearances, or which imports that this his appearance to his disciples in Galilee, in pursuance of his promise, was his first or only appearance, is made pretty evident by St. Mark's Gospel, which uses the same terms concerning the appearance in Galilee as St. Matthew uses, yet itself records two other appearances prior to this: 'Go your way, tell his disciples and Peter that he goeth before you into Galilee: there shall ye see him, as he said unto you' (xvi., 7). We might be apt to infer from these words, that this was the first time they were to see him: at least, we might infer it with as much reason as we draw the inference from the same words in Matthew; yet the historian himself did not perceive that he was leading his readers to any such conclusion, for in the twelfth and two following verses of this chapter, he informs us of two appearances, which, by comparing the order of events, are shown to have been prior to the appearance in Galilee. 'He appeared in another form unto two of them, as they walked, and went into the country: and they went and told it unto the residue: neither believed they them. Afterward He appeared unto the eleven as they sat at meat, and upbraided them with their unbelief, because they believed not them which had seen Him after He was risen.' Probably the same observation, concerning the particular design which guided the historian, may be of use in comparing many other passages of the Gospels."

[My brother's work, which has been interrupted by the letter and extract just given, will now be continued. What follows should be considered as coming immediately after the preceding chapter.—W. B. O.]

But there is a much worse set of notes than those on the twenty-eighth chapter of St. Matthew, and so important is it that we should put an end to such a style of argument, and get into a manner which shall commend itself to sincere and able adversaries, that I shall not apologise for giving them in full here. They refer to the spear wound recorded in St. John's Gospel as having been inflicted upon the body of our Lord.

The passage in St. John's Gospel stands thus (John xix., 32–37)—"Then came the soldiers and brake the legs of the first and of the other which was crucified with Him. But when they came to Jesus and saw that He was dead already they brake not His legs: but one of the soldiers with a spear pierced His side, and forthwith came there out blood and water. And he that saw it bare record, and we know that his record is true, and he knoweth that he saith true that ye might believe. For these things were done that the Scripture should be fulfilled, 'A bone of Him shall not be broken' and again another Scripture saith, 'They shall look on Him whom they pierced.'"

In his note upon the thirty-fourth verse Dean Alford writes—"The lance must have penetrated deep, for the object was to ensure death." Now what warrant is there for either of these assertions? We are told that the soldiers saw that our Lord was dead already, and that for this reason they did not break his legs: if there had been any doubt about His being dead can we believe that they would have hesitated? There is ample proof of the completeness of the death in the fact that those whose business it was to assure themselves of its having taken place were so satisfied that they would be at no further trouble; what need to kill a dead man? If there had been any question as to the possibility of life remaining, it would not have been resolved by the thrust of the spear, but in a way which we must shudder to think of. It is most painful to have had to write the foregoing lines, but are they not called for when we see a man so well intentioned and so widely read as the late Dean Alford condescending to argument which must only weaken the strength of his cause in the eyes of those who have not yet been brought to know the blessings and comfort of Christianity? From the words of St. John no one can say whether the wound was a deep one, or why it was given—yet the Dean continues, "and see John xx., 27," thereby implying that the wound must have been large enough for Thomas to get his hand into it, because our Lord says, "reach hither thine hand and thrust it into my side." This is simply shocking. Words cannot be pressed in this way. Dean Alford then says that the spear was thrust "probably into the left side on account of the position of the soldier" (no one can arrive at the position of the soldier, and no one would attempt to do so, unless actuated by a nervous anxiety to direct the spear into the heart of the Redeemer), "and of what followed" (the Dean here implies that the water must have come from the pericardium; yet in his next note we are led to infer that he rejects this supposition, inasmuch as the quantity of water would have been "so small as to have scarcely been observed"). Is this fair and manly argument, and can it have any other effect than to increase the scepticism of those who doubt?

Here this note ends. The next begins upon the words "blood and water."

"The spear," says the Dean, "perhaps pierced the pericardium or envelope of the heart" (but why introduce a "perhaps" when there is ample proof of the death without it?), "in which case a liquid answering to the description of water may have" (may have) "flowed with the blood, but the quantity would have been so small as scarcely to have been observed" (yet in the preceding note he has led us to suppose that he thinks the water "probably" came from near the heart). "It is scarcely possible that the separation of the blood into placenta and serum should have taken place so soon, or that if it had, it should have been described by an observe as blood and water. It is more probable that the fact here so strongly testified was a consequence of the extreme exhaustion of the body of the Redeemer." (Now if this is the case, the spear-wound does not prove the death of Him on whom it was inflicted, and Dean Alford has weakened a strong case for nothing.) "The medical opinions on the subject are very various and by no means satisfactory." Satisfactory! What does Dean Alford mean by satisfactory? If the evidence does not go to prove that the spear-wound must have been necessarily fatal why not have said so at once, and have let the whole matter rest in the obscurity from which no human being can remove it. The wound may have been severe or may not have been severe, it may have been given in mere wanton mockery of the dead King of the Jews, for the indignity's sake: or it may have been the savage thrust of an implacable foe, who would rejoice at the mutilation of the dead body of his enemy: none can say of what nature it was, nor why it was given; but the object of its having been recorded is no mystery, for we are expressly told that it was in order to shew that prophecy was thus fulfilled: the Evangelist tells us so in the plainest language: he even goes farther, for he says that these things were done for this end (not only that they were recorded)—so that the primary motive of the Almighty in causing the soldier to be inspired with a desire to inflict the wound is thus graciously vouchsafed to us, and we have no reason to harrow our feelings by supposing that a deeper thrust was given than would suffice for the fulfilment of the prophecy. May we not then well rest thankful with the knowledge which the Holy Spirit has seen fit to impart to us, without causing the weak brother to offend by our special pleading?

The reader has now seen the two first of Dean Alford's notes upon this subject, and I trust he will feel that I have used no greater plainness, and spoken with no greater severity than the case not only justifies but demands. We can hardly suppose that the Dean himself is not firmly convinced that our Lord died upon the Cross, but there are millions who are not convinced, and whose conviction should be the nearest wish of every Christian heart. How deeply, therefore, should we not grieve at meeting with a style of argument from the pen of one of our foremost champions, which can have no effect but that of making the sceptic suspect that the evidences for the death of our Lord are felt, even by Christians, to be insufficient. For this is what it comes to.

Let us, however, go on to the note on John xix., 35, that is to say on St. John's emphatic assertion of the truth of what he is recording. The note stands thus, "This emphatic assertion of the fact seems rather to regard the whole incident than the mere outflowing of the blood and water. It was the object of John to shew that the Lord's body was a real body and underwent real death." (This is not John's own account—supposing that John is the writer of the fourth Gospel—either of his own object in recording, or yet of the object of the wound's having been inflicted; his words, as we have seen above, run thus:—"and he that saw it bare record, and we know that his record is true; and he knoweth that he saith true that ye might believe. For these things were done that the Scripture should be fulfilled which saith 'a bone of him shall not be broken,' and, again, another Scripture saith, 'they shall look upon' him whom they pierced.'" Who shall dare to say that St. John had any other object than to show that the event which he relates had been long foreseen, and foretold by the words of the Almighty?) And both these were shewn by what took place, not so much by the phenomenon of the water and blood" (then here we have it admitted that so much disingenuousness has been resorted to for no advantage, inasmuch as the fact of the water and blood having flowed is not per se proof of a necessarily fatal wound) "as by the infliction of such a wound" (Such a wound! What can be the meaning of this? What has Dean Alford made clear about the wound? We know absolutely nothing about the severity or intention of the wound, and it is mere baseless conjecture and assumption to say that we do; neither do we know anything concerning its effect unless it be shewn that the issuing of the blood and water prove that death must have ensued, and this Dean Alford has just virtually admitted to be not shewn), after which, even if death had not taken place before (this is intolerable), there could not by any possibility be life remaining." (The italics on this page are mine.)

With this climax of presumptuous assertion these disgraceful notes are ended. They have shewn clearly that the wound does not in itself prove the death: they shew no less clearly that the Dean does not consider that the death is proved beyond possibility of doubt without the wound; what therefore should be the legitimate conclusion? Surely that we have no proof of the completeness of Christ's death upon the Cross—or in other words no proof of His having died at all! Couple this with the notes upon the Resurrection considered above, and we feel rather as though we were in the hands of some Jesuitical unbeliever, who was trying to undermine our faith in our most precious convictions under the guise of defending them, than in those of one whom it is almost impossible to suspect of such any design. What should we say if we had found Newton, Adam Smith or Darwin, arguing for their opinions thus? What should we think concerning any scientific cause which we found thus defended? We should exceedingly well know that it was lost. And yet our leading theologians are to be applauded and set in high places for condescending to such sharp practice as would be despised even by a disreputable attorney, as too transparently shallow to be of the smallest use to him.

After all that has been said either by Dean Alford or any one else, we know nothing more than what we are told by the Apostle, namely, that immediately before being taken down from the Cross our Lord's body was wounded more severely, or less severely, as the case may be, with the point of a spear, that from this wound there flowed something which to the eyes of the writer resembled blood and water, and that the whole was done in order that a well-known prophecy might be fulfilled. Yet his sentences in reference to this fact being ended, without his having added one iota to our knowledge upon the subject, the Dean gravely winds up by throwing a doubt upon the certainty of our Lord's death which was not felt by a single one of those upon the spot, and resting his clenching proof of its having taken place upon a wound, which he has just virtually admitted to have not been necessarily fatal. Nothing can be more deplorable either as morality or policy.

Yet the Dean is justified by the event. One would have thought he could have been guilty of nothing short of infatuation in hoping that the above notes would pass muster with any ordinarily intelligent person, but he knew that he might safely trust to the force of habit and prejudice in the minds of his readers, and his confidence has not been misplaced. Of all those engaged in the training of our young men for Holy Orders, of all our Bishops and clergy and tutors at colleges, whose very profession it is to be lovers of truth and candour, who are paid for being so, and who are mere shams and wolves in sheep's clothing if they are not ever on the look-out for falsehood, to make war upon it as the enemy of our souls—not one, no, not a single one, so far as I know, has raised his voice in protest. If a man has not lost his power of weeping let him weep for this; if there is any who realises the crime of self-deception, as perhaps the most subtle and hideous of all forms of sin, let him lift up his voice and proclaim it now; for the times are not of peace, but of a sowing of wind for the reaping of whirlwinds, and of the calm that is the centre of the hurricane.

Either Christianity is the truth of truths—the one which should in this world overmaster all others in the thoughts of all men, and compared with which all other truths are insignificant except as grouping themselves around it—or it is at the best a mistake which should be set right as soon as possible. There is no middle course. Either Jesus Christ was the Son of God, or He was not. If He was, His great Father forbid that we should juggle in order to prove Him so—that we should higgle for an inch of wound more, or an inch less, and haggle for the root νυγ in the Greek word ενυξε. Better admit that the death of Christ must be ever a matter of doubt, should so great a sacrifice be demanded of us, than go near to the handling of a lie in order to make assurance doubly sure. No truthful mind can doubt that the cause of Christ is far better served by exposing an insufficient argument than by silently passing it over, or else that the cause of Christ is one to be attacked and not defended.

CHAPTER VII

DIFFICULTIIES FELT BY OUR OPPONENTS

There are some who avoid all close examination into the circumstances attendant upon the death of our Lord, using the plea that however excellent a quality intellect may be, and however desirable that the facts connected with the Crucifixion should be intelligently considered, yet that after all it is spiritual insight which is wanted for a just appreciation of spiritual truths, and that the way to be preserved from error is to cultivate holiness and purity of life. This is well for those who are already satisfied with the evidences for their convictions. We could hardly give them any better advice than simply to "depart from evil, do good, seek peace and ensue it" (Psalm xxxiv., 14), if we could only make sure that their duty would never lead them into contact with those who hold the external evidences of Christianity to be insufficient. When, however, they meet with any of these unhappy persons they will find their influence for good paralysed; for unbelievers do not understand what is meant by appealing to their spiritual insight as a thing which can in any way affect the evidence for or against an alleged fact in history—or at any rate as forming evidence for a fact which they believe to be in itself improbable and unsupported by external proof. They have not got any spiritual insight in matters of this sort; nor, indeed, do they recognise what is meant by the words at all, unless they be interpreted as self-respect and regard for the feelings and usages of other people. What spiritual insight they have, they express by the very nearly synonymous terms, "current feeling," or "common sense," and however deep their reverence for these things may be, they will never admit that goodness or right feeling can guide them into intuitive accuracy upon a matter of history. On the contrary, in any such case they believe that sentiment is likely to mislead, and that the well-disciplined intellect is alone trustworthy. The question is, whether it is worth while to try and rescue those who are in this condition or not. If it is worth while, we must deal with them according to their sense of right and not ours: in other words, if we meet with an unbeliever we must not expect him to accept our faith unless we take much pains with him, and are prepared to make great sacrifice of our own peace and patience.

Yet how many shrink from this, and think that they are doing God service by shrinking; the only thing from which they should really shrink, is the falsehood which has overlaid the best established fact in all history with so much sophistry, that even our own side has come to fear that there must be something lurking behind which will not bear daylight; to such a pass have we been brought by the desire to prove too much.

Now for the comfort of those who may feel an uneasy sense of dread, as though any close examination of the events connected with the Crucifixion might end in suggesting a natural instead of a miraculous explanation of the Resurrection, for the comfort of such—and they indeed stand in need of comfort—let me say at once that the ablest of our adversaries would tell them that they need be under no such fear. Strauss himself admits that our Lord died upon the Cross; he does not even attempt to dispute it, but writes as though he were well aware that there was no room for any difference of opinion about the matter. He has therefore been compelled to adopt the hallucination theory, with a result which we have already considered. Yet who can question that Strauss would have maintained the position that our Lord did not die upon the Cross, unless he had felt that it was one in which he would not be able to secure the support even of those who were inclined to disbelieve? We cannot doubt that the conviction of the reality of our Lord's death has been forced upon him by a weight of testimony which, like St. Paul, he has found himself utterly unable to resist.

Here then, we might almost pause. Strauss admits that our Lord died upon the Cross. Yet can the reader help feeling that the vindication of the reality of our Lord's reappearances, and the refutation of Strauss's theories with which this work opened, was triumphant and conclusive? Then what follows? That Christ died and rose again! The central fact of our faith is proved. It is proved externally by the most solid and irrefragable proofs, such as should appeal even to minds which reject all spiritual evidence, and recognise no canons of investigation but those of the purest reason.

But anything and everything is believable concerning one whose resurrection from death to life has been established. What need, then, to enter upon any consideration of the other miracles? Of the Ascension? Of the descent of the Holy Spirit? Who can feel difficulty about these things? Would not the miracle rather be that they should not have happened! May we not now let the wings of our soul expand, and soar into the heaven of heavens, to the footstool of the Throne of Grace, secure that we have earned the right to hope and to glory by having consented to the pain of understanding?

We may: and I have given the reader this foretaste of the prize which he may justly claim, lest he should be swallowed up in overmuch grief at the journey which is yet before him ere he shall have done all which may justly be required of him. For it is not enough that his own sense of security should be perfected. This is well; but let him also think of others.

What then is their main difficulty, now that it has been shewn that the reappearances of our Lord were not due to hallucination?

I propose to shew this by collecting from all the sources with which I was familiar in former years, and throwing the whole together as if it were my own. I shall spare no pains to make the argument tell with as much force as fairness will allow. I shall be compelled to be very brief, but the unbeliever will not, I hope, feel that anything of importance to his side has been passed over. The believer, on the other hand, will be thankful both to know the worst and to see how shallow and impotent it will appear when it comes to be tested. Oh! that this had been done at the beginning of the controversy, instead of (as I heartily trust) at the end of it.

Our opponents, therefore, may be supposed to speak somewhat after the following manner:—
"Granted," they will say, "for the sake of argument, that Jesus Christ did reappear alive after his Crucifixion; it does not follow that we should at once necessarily admit that his reappearance was due to miracle. What was enough, and reasonably enough, to make the first Christians accept the Resurrection, and hence the other miracles of Christ, is not enough and ought not to be enough to make men do so now. If we were to hear now of the reappearance of a man who had been believed to be dead, our first impulse would be to learn the when and where of the death, and the when and where of the first reappearance. What had been the nature of the death? What conclusive proof was there that the death had been actual and complete? What examination had been made of the body? And to whom had it been delivered on the completeness of the death having been established? How long had the body been in the grave—if buried? What was the condition of the grave on its being first revisited? It is plain to any one that at the present day we should ask the above questions with the most jealous scrutiny and that our opinion of the character of the reappearance would depend upon the answers which could be given to them.

"But it is no less plain that the distance of the supposed event from our own time and country is no bar to the necessity for the same questions being as jealously asked concerning it, as would be asked if it were alleged to have happened recently and nearer home. On the contrary, distance of time and space introduces an additional necessity for caution. It is one thing to know that the first Christians unanimously believed that their master had miraculously risen from death to life; it is another to know their reasons for so thinking. Times have changed, and tests of truth are infinitely better understood, so that the reasonable of those days is reasonable to us no longer. Nor would it be enough that the answers given could be just strained into so much agreement with one another as to allow of a modus vivendi between them, and not to exclude the possibility of death, they must exclude all possibility of life having remained, or we should not hesitate for a moment about refusing to believe that the reappearance had been miraculous: indeed, so long as any chink or cranny or loophole for escape from the miraculous was afforded to us, we should unhesitatingly escape by it; this, at least, is the course which would be adopted by any judge and jury of sensible men if such a case were to come before their unprejudiced minds in the common course of affairs.

"We should not refuse to believe in a miracle even now, if it were supported by such evidence as was considered to be conclusive by the bench of judges and by the leading scientific men of the day: in such a case as this we should feel bound to accept it; but we cannot believe in a miracle, no matter how deeply it has been engrained into the creeds of the civilised world, merely because it was believed by 'unlettered fishermen' two thousand years ago. This is not a source from which such an event as a miracle should be received without the closest investigation. We know, indeed, that the Apostles were sincere men, and that they firmly believed that Jesus Christ had risen from the dead; their lives prove their faith; but we cannot forget that the fact itself of Christ's having been crucified and afterwards seen alive, would be enough, under the circumstances, to incline the men of that day to believe that he had died and had been miraculously restored to life, although we should ourselves be bound to make a far more searching inquiry before we could arrive at any such conclusion. A miracle was not and could not be to them, what it is and ought to be to ourselves—a matter to be regarded a priori with the very gravest suspicion. To them it was what it is now to the lower and more ignorant classes of Irish, French, Spanish and Italian peasants: that is to say, a thing which was always more or less likely to happen, and which hardly demanded more than a primâ facie case in order to establish its credibility. If we would know what the Apostles felt concerning a miracle, we must ask ourselves how the more ignorant peasants of to-day feel: if we do this we shall have to admit that a miracle might have been accepted upon very insufficient grounds, and that, once accepted, it would not have had one-hundredth part so good a chance of being refuted as it would have now.

"It should be borne in mind, and is too often lost sight of, that we have no account of the Resurrection from any source whatever. We have accounts of the visit of certain women to a tomb which they found empty; but this is not an account of a resurrection. We are told that Jesus Christ was seen alive after being thought to have been dead, but this again is not an account of a resurrection. It is a statement of a fact, but it is not an account of the circumstances which attended that fact. In the story told by Matthew we have what comes nearest to an account of the Resurrection, but even here the principal figure is wanting; the angel rolls away the stone and sits upon it, but we hear nothing about the body of Christ emerging from the tomb; we only meet with this, when we come to the Italian painters.

"Moreover, St. Matthew's account is utterly incredible from first to last; we are therefore thrown back upon the other three Evangelists, none of whom professes to give us the smallest information as to the time and manner of Christ's Resurrection. There is nothing in any of their accounts to preclude his having risen within two hours from his having been laid in the tomb.

"If a man of note were condemned to death, crucified and afterwards seen alive, the almost instantaneous conclusion in the days of the Apostles, and in such minds as theirs, would be that he had risen from the dead; but the almost instantaneous conclusion now, among all whose judgement would carry the smallest weight, would be that he had never died—that there must have been some mistake. Children and inexperienced persons believe readily in all manner of improbabilities and impossibilities, which when they become older and wiser they cannot conceive their having ever seriously accepted. As with men, so with ages; an unusual train of events brings about unusual results, whereon the childlike age turns instinctively to miracle for a solution of the difficulty. In the days of Christ men would ask for evidence of the Crucifixion and the reappearance; when these two points had been established they would have been satisfied—not unnaturally—that a great miracle had been performed: but no sane man would be contented now with the evidence that was sufficient then, any more than he would be content to accept many things which a child must take upon authority, and authority only. We ought to require the most ample evidence that not only the appearance of death, but death itself, must have inevitably ensued upon the Crucifixion, and if this were not forthcoming we should not for a moment hesitate about refusing to believe that the reappearance was miraculous.

"And this is what would most assuredly be done now by impartial examiners—by men of scientific mind who had no wish either to believe or disbelieve except according to the evidence; but even now, if their affections and their hopes of a glorious kingdom in a world beyond the grave were enlisted on the side of the miracle, it would go hard with the judgement of most men. How much more would this be so, if they had believed from earliest childhood that miracles were still occasionally worked in England, and that a few generations ago they had been much more signal and common?

"Can we wonder then, if we ourselves feel so strongly concerning events which are hull down upon the horizon of time, that those who lived in the very thick of them should have been possessed with an all absorbing ecstasy or even frenzy of excitement? Assuredly there is no blame on the score of credulity to be attached to those who propagated the Christian religion, but the beliefs which were natural and lawful to them, are, if natural, yet not lawful to ourselves: they should be resisted: they are neither right nor wise, and do not form any legitimate ground for faith: if faith means only the believing facts of history upon insufficient evidence, we deny the merit of faith; on the contrary, we regard it as one of the most deplorable of all errors—as sapping the foundations of all the moral and intellectual faculties. It is grossly immoral to violate one's inner sense of truth by assenting to things which, though they may appear to be supported by much, are still not supported by enough. The man who can knowingly submit to such a derogation from the rights of his self-respect, deserves the injury to his mental eye-sight which such a course will surely bring with it. But the mischief will unfortunately not be confined to himself; it will devolve upon all who are ill-fated enough to be in his power; he will be reckless of the harm he works them, provided he can keep its consequences from being immediately offensive to himself. No: if a good thing can be believed legitimately, let us believe it and be thankful, otherwise the goodness will have departed out of it; it is no longer ours; we have no right to it, and shall suffer for it, we and our children, if we try to keep it. It has been said that the fathers have eaten sour grapes, and the children's teeth are set on edge, but, more truly, it is the eating of sweet and stolen fruit by the fathers that sets the teeth of the children jarring. Let those who love their children look to this, for on their own account they may be mainly trusted to avoid the sour. Hitherto the intensity of the belief of the Apostles has been the mainstay of our own belief. But that mainstay is now no longer strong enough. A rehearing of the evidence is imperatively demanded, that it may either be confirmed or overthrown."

It cannot be denied that there is much in the above with which all true Christians will agree, and little to find fault with except the self-complacency which would seem to imply that common sense and plain dealing belong exclusively to the unbelieving side. It is time that this spirit should be protested against not in word only but in deed. The fact is, that both we and our opponents are agreed that nothing should be believed unless it can be proved to be true. We repudiate the idea that faith means the accepting historical facts upon evidence which is insufficient to establish them. We do not call this faith; we call it credulity, and oppose it to the utmost of our power.

Our opponents imply that we regard as a virtue well-pleasing in the sight of God, and dignify with the name of faith, a state of mind which turns out to be nothing but a willingness to stand by all sorts of wildly improbable stories which have reached us from a remote age and country, and which, if true, must lead us to think otherwise of the whole course of nature than we should think if we were left to ourselves. This accusation is utterly false and groundless. Faith is the "evidence of things not seen," but it is not "insufficient evidence for things alleged to have been seen." It is "the substance of things hoped for," but "reasonably hoped for" was unquestionably intended by the Apostle. We base our faith in the deeper mysteries of our religion, as in the nature of the Trinity and the sacramental graces, upon the certainty that other things which are within the grasp of our reason can be shewn to be beyond dispute. We know that Christ died and rose again; therefore we believe whatever He sees fit to tell us, and follow Him, or endeavour to follow Him, whereinsoever He commands us, but we are not required to take both the commands of the Mediator and His credentials upon faith. It is because certain things within our comprehension are capable of the most irrefragable proof, that certain others out of it may justly be required to be believed, and indeed cannot be disbelieved without contumacy and presumption. And this applies to a certain extent to the credentials also: for although no man should be captious, nor ask for more evidence than would satisfy a well-disciplined mind concerning the truth of any ordinary fact (as one who not contented with the evidence of a seal, a handwriting and a matter not at variance with probability, would nevertheless refuse to act upon instructions because he had not with his own eyes actually seen the sender write and sign and seal), yet it is both reasonable and indeed necessary that a certain amount of care should be taken before the credentials are accepted. If our opponents mean no more than this we are at one with them, and may allow them to proceed.

"Turn then," they say, "to the account of the events which are alleged to have happened upon the morning of the Resurrection, as given in the fourth Gospel: and assume for the sake of the argument that that account, if not from John's own hand, is nevertheless from a Johannean source, and virtually the work of the Apostle. The account runs as follows:

"'The first day of the week cometh Mary Magdalene while it was yet dark unto the sepulchre, and seeth the stone taken away from the sepulchre. Then she runneth and cometh to Simon Peter and to the other disciple whom Jesus loved, and saith unto them, 'They have taken away the Lord out of the sepulchre, and we know not where they have laid Him.' Peter therefore went forth and that other disciple, and came to the sepulchre. So they both ran together: and the other disciple did outrun Peter, and came first to the sepulchre. And he stooping down and looking in, saw the linen clothes lying, yet went he not in. Then cometh Simon Peter following him and went into the sepulchre and seeth the linen clothes lie, and the napkin that was about His head not lying with the linen clothes but wrapped together in a place by itself. Then went in also that other disciple, which came first to the sepulchre, and he saw and believed. For as yet they knew not the Scripture that he must rise from the dead. Then the disciples went away again to their own home. But Mary stood without at the sepulchre weeping; and as she wept, she stooped down, and looked into the sepulchre, and seeth two angels in white sitting, the one at the head, the other at the feet, where the body of Jesus had lain, and they say unto her, 'Woman, why weepest thou?' She saith unto them, 'Because they have taken away my Lord and I know not where they have laid him.'"

"Then Mary sees Jesus himself, but does not at first recognise him.

"Now, let us see what the above amounts to, and, dividing it into two parts, let us examine first what we are told as having come actually under John's own observation, and, secondly, what happened afterwards.

I. "It is clear that Mary had seen nothing miraculous before she came running to the two Apostles, Peter and John. She had found the tomb empty when she reached it. She did not know where the body of her Lord then was, nor was there anything to shew how long it had been removed: all she knew was that within thirty-six hours from the time of its having been laid in the tomb it had disappeared, but how much earlier it had been gone neither did she know, nor shall we. Peter and John went into the sepulchre and thoroughly examined it: they saw no angel, nor anything approaching to the miraculous, simply the grave clothes (which were probably of white linen), lying in two separate places. Then, and not till then, do they appear to have entertained their first belief or hope that Christ might have risen from the dead.

"This is plain and credible; but it amounts to an empty tomb, and to an empty tomb only.

"Here, for a moment, we must pause. Had these men but a few weeks previously seen Lazarus raised from the corruption of the grave—to say nothing of other resurrections from the dead? Had they seen their master override every known natural law, and prove that, as far as he was concerned, all human experience was worthless, by walking upon rough water, by actually talking to a storm of wind and making it listen to him, by feeding thousands with a few loaves, and causing the fragments that remained after all had eaten, to be more than the food originally provided? Had they seen events of this kind continually happening for a space of some two years, and finally had they seen their master transfigured, conversing with the greatest of their prophets (men who had been dead for ages), and recognised by a voice from heaven as the Son of the Almighty, and had they also heard anything approaching to an announcement that he should himself rise from the dead—or had they not? They might have seen the raising of Lazarus and the rest of the miracles, but might not have anticipated that Christ himself would rise, for want of any announcement that this should be so; or, again, they might have heard a prophecy of his Resurrection from the lips of Christ, but disbelieved it for the want of any previous miracles which should convince them that the prophecy came from no ordinary person; so that their not having expected the Resurrection is explicable by giving up either the prophecies, or the miracles, but it is impossible to believe that in spite both of the miracles and the prophecies, the Apostles should have been still without any expectation of the Resurrection. If they had both seen the miracles and heard the prophecies, they must have been in a state of inconceivably agitated excitement in anticipation of their master's reappearance. And this they were not; on the contrary, they were expecting nothing of the kind. The condition of mind ascribed to them considering their supposed surroundings, is one which belongs to the drama only; it is not of nature: it is so utterly at variance with all human experience that it should be dismissed at once as incredible.

"But it is very credible if Christ was seen alive after his Crucifixion, and his reappearance, though due to natural causes, was once believed to be miraculous, that this one seemingly well substantiated miracle should become the parent of all the others, and of the prophecies of the Resurrection. Thirty years in all probability elapsed between the reappearances of Christ and the earliest of the four Gospels; thirty years of oral communication and spiritual enthusiasm, among an oriental people, and in an unscientific age; an age by which the idea of an interference with the modes of the universe from a point outside of itself, was taken as a matter of course; an age which believed in an anthropomorphic Deity who had back parts, which Moses had been allowed to see through the hand of God; an age which, over and above all this, was at the time especially convulsed with expectations of deliverance from the Roman yoke. Have we not here a soil suitable for the growth of miracles, if the seed once fell upon it? Under such conditions they would even spring up of themselves, seedless.

"Once let the reappearances of Christ have been believed to be miraculous (and under all the circumstances they might easily have been believed to be so, though due to natural causes), and it is not wonderful that, in such an age and among such a people, the other miracles and the prophecies of the Resurrection should have become current within thirty years. Even we ourselves, with all our incalculably greater advantages, could not withstand so great a temptation to let our wish become father to our thoughts. If we had been the especially favoured friends of one whom we believed to have died, but who yet was not to beholden by death, no matter how careful and judicially minded we might be by nature, we should be blind to everything except the fact that we had once been the chosen companions of an immortal. There lives no one who could withstand the intoxication of such an idea. A single well-substantiated miracle in the present day, even though we had not seen it ourselves, would uproot the hedges of our caution; it would rob us of that sense of the continuity of nature, in which our judgements are, consciously or unconsciously, anchored; but if we were very closely connected with it in our own persons, we should dwell upon the recollection of it and on little else.

"Few of us can realise what happened so very long ago. Men believe in the Christian miracles, though they would reject the notion of a modern miracle almost with ridicule, and would hardly even examine the evidence in its favour. But the Christian miracles stand in their minds as things apart; their prestige is greater than that attaching to any other events in the whole history of mankind. They are hallowed by the unhesitating belief of many, many generations. Every circumstance which should induce us to bow to their authority surrounds them with a bulwark of defences which may make us well believe that they must be impregnable, and sacred from attack. Small wonder then that the many should still believe them. Nevertheless they do not believe them so fully, nor nearly so fully, as they think they do. For even the strongest imagination can travel but a very little way beyond a man's own experience; it will not bear the burden of carrying him to a remote age and country; it will flag, wander and dream; it will not answer truly, but will lay hold of the most obvious absurdity, and present it impudently to its tired master, who will accept it gladly and have done with it. Even recollection fails, but how much more imagination! It is a high flight of imagination to be able to realise how weak imagination is.

"We cannot therefore judge what would be the effect of immediate contact even with the wild hope of a miracle, from our conventional acceptance of the Christian miracles. If we would realise this we must look to modern alleged miracles—to the enthusiasm of the Irish and American revivals, when mind inflames mind till strong men burst into hysterical tears like children; we must look for it in the effect produced by the supposed Irvingite miracles on those who believed in them, or in the miracles that followed the Port Royal miracle of the holy thorn. There never was a miracle solitary yet: one will soon become the parent of many. The minds of those who have believed in a single miracle as having come within their own experience become ecstatic; so deeply impressed are they with the momentous character of what they have known, that their power of enlisting sympathy becomes immeasurably greater than that of men who have never believed themselves to have come into contact with the miraculous; their deep conviction carries others along with it, and so the belief is strengthened till adverse influences check it, or till it reaches a pitch of grotesque horror, as in the case of the later Jansenist miracles. There is nothing, therefore, extraordinary in the gradual development within thirty years of all the Christian miracles, if the Resurrection were once held to be well substantiated; and there is nothing wonderful, under the circumstances, in the reappearance of Christ alive after his Crucifixion having been assigned to miracle. He had already made sufficient impression upon his followers to require but little help from circumstances. He had not so impressed them as to want no help from any supposed miracle, but nevertheless any strange event in connection with him would pass muster, with little or no examination, as being miraculous. He had undoubtedly professed himself to be, and had been half accepted as, the promised Messiah. He had no less undoubtedly appeared to be dead, and had been believed to be so both by friends and foes. Let us also grant that he reappeared alive. Would it, then, be very astonishing that the little missing link in the completeness of the chain of evidence—absolute certainty concerning the actuality of the death—should have been allowed to drop out of sight?

"Round such a centre, and in such an age, the other miracles would spring up spontaneously, and be accepted the moment that they arose; there is nothing in this which is foreign to the known tendencies of the human mind, but there would be something utterly foreign to all we know of human nature, in the fact of men not anticipating that Christ would rise, if they had already seen him raise others from the dead and work the miracles ascribed to him, and if they had also heard him prophesy that he should himself rise from the dead. In fact nothing can explain the universally recorded incredulity of the Apostles as to the reappearance of Christ, except the fact that they had never seen him work a single miracle, or else that they had never heard him say anything which could lead them to suppose that he was to rise from the dead.

"We are therefore not unwilling to accept the facts recorded in the fourth Gospel, in so far as they inform us of things which came under the knowledge of the writer. Mary found the tomb empty. Ignorant alike of what had taken place and of what was going to happen, she came to Peter and John to tell them that the body was gone; this was all she knew. The two go to the tomb, and find all as Mary had said; on this it is not impossible that a wild dream of hope may have flashed upon their minds, that the aspirations which they had already indulged in were to prove well founded. Within an hour or two Christ was seen alive, nor can we wonder if the years which intervened between the morning of the Resurrection and the writing of the fourth Gospel, should have sufficed to make the writer believe that John had had an actual belief in the Resurrection, while in truth he had only wildly hoped it. This much is at any rate plain, that neither he nor Peter had as yet heard any clearly intelligible prophecy that their master should rise from the dead. Whatever subsequent interpretation may have been given to some of the sayings of Jesus Christ, no saying was yet known which would of itself have suggested any such inference. We may justly doubt the caution and accuracy of the first founders of Christianity, without, even in our hearts, for one moment impugning the honesty of their intentions. We are ready to admit that had we been in their places we should in all likelihood have felt, believed, and, we will hope, acted as they did; but we cannot and will not admit, in the face of so much evidence to the contrary, that they were superior to the intelligence of their times, or, in other words, that they were capable critics of an event, in which both their feelings and the primâ facie view of the facts would be so likely to mislead them.

II. "Turning now to the narrative of what passed when Peter and John were gone, we find that Mary, stooping down, looked through her tears into the darkness of the tomb, and saw two angels clothed in white, who asked her why she wept. We must remember the wide difference between believing what the writer of the fourth Gospel tells us that John saw, and what he tells us that Mary Magdalene saw. All we know on this point is that he believed that Mary had spoken truly. Peter and John were men, they went into the tomb itself, and we may say for a certainty that they saw no angel, nor indeed anything at all, but the grave clothes (which were probably of white linen), lying in two separate places within it. Mary was a woman—a woman whose parallel we must look for among Spanish or Italian women of the lower orders at the present day; she had, we are elsewhere told, been at one time possessed with devils; she was in a state of tearful excitement, and looking through her tears from light into comparative darkness. Is it possible not to remember what Peter and John did see when they were in the tomb? Is it possible not to surmise that Mary in good truth saw nothing more? She thought she saw more, but the excitement under which she was labouring at the time, an excitement which would increase tenfold after she had seen Christ (as she did immediately afterwards and before she had had time to tell her story), would easily distort either her vision or her memory, or both.

"The evidence of women of her class—especially when they are highly excited—is not to be relied upon in a matter of such importance and difficulty as a miracle. Who would dare to insist upon such evidence now? And why should it be considered as any more trustworthy eighteen hundred years ago? We are indeed told that the angels spoke to her; but the speech was very short; the angels simply ask her why she weeps; she answers them as though it were the common question of common people, and then leaves them. This is in itself incredible; but it is not incredible that if Mary looking into the tomb saw two white objects within, she should have drawn back affrighted, and that her imagination, thrown into a fever by her subsequent interview with Christ, should have rendered her utterly incapable of recollecting the true facts of the case; or, again, it is not incredible that she should have been believed to have seen things which she never did see. All we can say for certain is that before the fourth Gospel was written, and probably shortly after the first reappearance of Christ, Mary Magdalene believed, or was thought to have believed, that she had seen angels in the tomb; and this being so, the development of the short and pointless question attributed to them—possibly as much due to the eager cross-questioning of others as to Mary herself—is not surprising.

"Before the Sunday of the Resurrection was over, the facts as derivable from the fourth Gospel would stand thus. Jesus Christ, who was supposed to have been verily and indeed dead, was known to be alive again. He had been seen, and heard to speak. He had been seen by those who were already prepared to accept him as their leader, and whose previous education, and tone of mind, would lead them rather to an excess of faith in a miracle, than of scepticism concerning its miraculous character. The Apostles would be in no impartial nor sceptical mood when they saw that Christ was alive. The miracle was too near themselves—too fascinating in its supposed consequences for themselves—to allow of their going into curious questions about the completeness of the death. The Master whom they had loved, and in whom they had hoped, had been crucified and was alive again. Is it a harsh or strained supposition, that what would have assuredly been enough for ourselves, if we had known and loved Christ and had been attuned in mind as the Apostles were, should also have been enough for them? Who can say so? The nature of our belief in our Master would have been changed once and for ever; and so we find it to have been with the Christian Apostles.

"Over and above the reappearance of Christ, there would also be a report (probably current upon the very Sunday of the Resurrection), that Mary Magdalene had seen a vision of angels in the tomb in which Christ's body had been laid; and this, though a matter of small moment in comparison with the reappearance of Christ himself, will nevertheless concern us nearly when we come to consider the narratives of the other Evangelists."

CHAPTER VIII

THE PRECEDING CHAPTER (continued)

"Let us now turn to Luke. His account runs as follows:—

"'Now upon the first day of the week, very early in the morning, they came unto the sepulchre bringing the spices which they had prepared, and certain others with them. And they found the stone rolled away from the sepulchre. And they entered in, and found not the body of the Lord Jesus. And it came to pass as they were much perplexed thereabout, behold, two men stood by them in shining garments, and as they were afraid, and bowed their faces to the earth, they said unto them, "Why seek ye the living among the dead? He is not here, but is risen: remember how he spake unto you when he was yet in Galilee, saying, 'The Son of Man must be delivered into the hands of sinful men and be crucified, and the third day rise again." And they remembered his words, and returned from the sepulchre, and told all these things unto the eleven, and to all the rest. It was Mary Magdalene and Joanna, and Mary the mother of James, and other women that were with them which told these things unto the Apostles. And their words seemed unto them as idle tales, and they believed them not. Then arose Peter, and went unto the sepulchre: and, stooping down, he beheld the linen clothes laid by themselves, and departed wondering in himself at that which was come to pass.'

"When we compare this account with John's we are at once struck with the resemblances and the discrepancies. Luke and John indeed are both agreed that Christ was seen alive after the Crucifixion. Both agree that the tomb was found empty very early on the Sunday morning (i.e., within thirty-six hours of the deposition from the Cross), and neither writer affords us any clue whatever as to the time and manner of the removal of the body; but here the resemblances end; the angelic vision of Mary, seen after Peter and John had departed from the tomb, and seen apparently by Mary alone, in Luke finds its way into the van of the narrative, and Peter is represented as having gone to the tomb, not in consequence of having been simply told that the body of Christ was missing, but because he refused to believe the miraculous story which was told him by the women. In the fourth Gospel we heard of no miraculous story being carried by Mary to Peter and John. The angels instead of being seen by one person only, as would have appeared from the fourth Gospel, are now seen by many; and the women instead of being almost stolidly indifferent to the presence of supernatural beings, are afraid, and bow down their faces to the earth; instead of merely wanting to be informed why Mary was weeping, the angels speak with definite point, and as angels might be expected to speak; they allude, also, to past prophecy, which the women at once remember.

"Strange, that they should want reminding! And stranger still that a few verses lower down we should find the Apostles remembering no prophetic saying, but regarding the story of the women as mere idle tales. What shall we say? Are not these differences precisely similar to those which we are continually meeting with, when a case of exaggeration comes before us? Can we accept both the stories? Is this one of those cases in which all would be made clear if we did but know all the facts, or is it rather one in which we can understand how easily the story given by the one writer might become distorted into the version of the other? Does it seem in any way improbable that within the forty years or so between the occurrences recorded by John and the writing of Luke's Gospel, the apparently trifling, yet truly most important, differences between the two writers should have been developed?

"No one will venture to say that the facts, upon the face of them, do not strongly suggest such an inference, and that, too, with no conscious fraud on the part of any of those through whose mouths the story must have passed. If the fourth Gospel be assigned to John (and if it is not assigned to John the difficulties on the Christian side become so great that the cause may be declared lost), his story is that of a principal actor and eye-witness; it bears every impress of truth and none of exaggeration upon any point which came under his own observation. Even when he tells of what Mary Magdalene said she saw, we see the myth in its earliest and crudest form; there is no attempt at circumstance in connection with it, and abundant reason for suspecting its supernatural character is given along with it; reason which to our minds is at any rate sufficient to make us doubt it, but which would naturally have no weight whatever with John after he had once seen Christ alive, or indeed with us if we had been in his place. It is not to be wondered at that in such times many a fresh bud should be grafted on to the original story; indeed it was simply inevitable that this should have been the case. No one would mean to deceive, but we know how, among uneducated and enthusiastic persons, the marvellous has an irresistible tendency to become more marvellous still; and, as far as we can gather, all the causes which bring this about were more actively at work shortly after the time of Christ's first reappearance than at any other time which can be readily called to mind. The main facts, as we derive them from the consent of both writers, were simply these:—That the tomb of Christ was found unexpectedly empty on the Sunday morning; that this fact was reported to the Apostles; that Peter went into the tomb and saw the linen clothes laid by themselves; that Mary Magdalene said that she had seen angels; and that eventually Christ shewed himself undoubtedly alive. Both writers agree so far, but it is impossible to say that they agree farther.

"Some may say that it is of little moment whether the angels appeared first or last; whether they were seen by many or by one; whether, if seen only by one, that one had previously been insane; whether they spoke as angels might be expected to speak, i.e., to the point, and are shewn to have been recognised as angels by the fear which their appearance caused; or whether they caused no alarm, and said nothing which was in the least equal to the occasion. But most men will feel that the whole complexion of the story changes according to the answers which can be made to these very questions. Surely they will also begin to feel a strong suspicion that the story told by Luke is one which has not lost in the telling. How natural was it that the angelic vision should find its way into the foreground of the picture, and receive those little circumstantial details of which it appeared most to stand in need; how desirable also that the testimony of Mary should be corroborated by that of others who were with her, and out of whom no devils had been cast. The first Christians would not have been men and women at all unless they had felt thus; but they were men and women, and hence they acted after the fashion of their age and unconsciously exaggerated; the only wonder is that they did not exaggerate more, for we must remember that even though the Apostles themselves be supposed to have been more judicially unimpassioned and less liable to inaccuracy than we have reason to believe they were, yet that from the very earliest ages of the Church there would be some converts of an inferior stamp. No matter how small a society is, there will be bad in it as well as good—there was a Judas even in the twelve.

"But to speak less harshly, there must from the first have been some converts who would be capable of reporting incautiously; visions and dreams were vouchsafed to many, and not a few marvels may be referable to this source; there is no trusting an age in which men are liable to give a supernatural interpretation to an extraordinary dream, nor is there any end to what may come of it, if people begin seriously confounding their sleeping and waking impressions. In such times, then, Luke may have said with a clear conscience that he had carefully sifted the truth of what he wrote; but the world has not passed through the last two thousand years in vain, and we are bound to insist upon a higher standard of credibility. Luke would believe at once, and as a matter of course, things which we should as a matter of course reject; yet it is probable that he too had heard much that he rejected; he seems to have been dissatisfied with all the records with the existence of which he was aware; the account which he gives is possibly derived from some very early report; even if this report arose at Jerusalem, and within a week after the Crucifixion, it might well be very inaccurate, though apparently supported by excellent authority, so that there is no necessity for charging Luke with unusual credulity. No one can be expected to be greatly in advance of his surroundings; it is well for every one except himself if he should happen to be so, but no man is to be blamed if he is not; it is enough to save him if he is fairly up to the standard of his own times. 'Morality' is rather of the custom which is, than of the custom which ought to be.

"Turning now to the account of Mark, we find the following:—

"'And when the Sabbath was past, Mary Magdalene, and Mary the mother of James, and Salome had bought sweet spices that they might come and anoint him. And very early in the morning, the first day of the week, they came unto the sepulchre at the rising of the sun. And they said among themselves,

"Who shall roll us away the stone from the door of the sepulchre?" And when they looked they saw that the stone was rolled away; for it was very great. And entering into the sepulchre they saw a young man sitting on the right side, clothed in a long white garment; and they were affrighted. And he saith unto them, "Be not affrighted; ye seek Jesus of Nazareth which was crucified; he is risen; he is not here; behold the place where they laid him. But go your way, tell his disciples and Peter that he goeth before you into Galilee: there ye shall see him, as he said unto you." And they went out quickly, and fled from the sepulchre; for they trembled and were amazed, neither said they any thing to any man, for they were afraid. Now when Jesus was risen early the first day of the week, he appeared first to Mary Magdalene, out of whom he had cast seven devils. And she went and told them that had been with him as they mourned and wept. And they, when they heard that he was alive, and had been seen of her, believed not.'

"Here we have substantially the same version as that given by Luke; there is only one angel mentioned, but it may be said that it is possible that there may have been another who is not mentioned, inasmuch as he remained silent; the angelic vision, however, is again brought into the foreground of the story and the fear of the women is even more strongly insisted on than it was in Luke. The angel reminds the women that Christ had said that he should be seen by his Apostles in Galilee, of which saying we again find that the Apostles seem to have had no recollection. The linen clothes have quite dropped out of the story, and we can detect no trace of Peter and John's visit to the tomb, but it is remarkable that the women are represented as not having said anything about the presence of the angel immediately on their having seen him; and this fact, which might be in itself suspicious, is apologised for on the score of fear, notwithstanding that their silence was a direct violation of the command of the being whom they so greatly feared. We should have expected that if they had feared him so much they would have done as he told them, but here again everybody seems to act as in a dream or drama, in defiance of all the ordinary principles of human action.

"Throughout the preceding paragraph we have assumed that Mark intended his readers to understand that the young man seen in the tomb was an angel; but, after all, this is rather a bold assumption. On what grounds is it supported? Because Luke tells us that when the women reached the tomb they found two white angels within it, are we therefore to conclude that Mark, who wrote many years earlier, and as far as we can gather with much greater historical accuracy, must have meant an angel when he spoke of a 'young man'? Yet this can be the only reason, unless the young man's having worn a long white robe is considered as sufficient cause for believing him to have been an angel; and this, again, is rather a bold assumption. But if St. Mark meant no more than he said, and when he wrote of a 'young man' intended to convey the idea of a young man and of nothing more, what becomes of the angelic visions at the tomb of Christ? For St. Matthew's account is wholly untenable; St. Luke is a much later writer, who must have got all his materials second or third hand; and although we granted, and are inclined to believe, that the accounts of the visits of Mary Magdalene, and subsequently of Peter and John to the tomb, which are given in the fourth Gospel, are from a Johannean source, if we were asked our reasons for this belief, we should be very hard put to it to give them. Nevertheless we think it probable.

"But take it either way; if the account in the fourth Gospel is supposed to have been derived from the Apostle John, we have already seen that there is nothing miraculous about it, so far as it deals with what came under John's own observation; if, on the other hand, it is not authentic we are thrown back upon St. Mark as incomparably our best authority for the facts that occurred on the Sunday after the Crucifixion, and he tells us of nothing but a tomb found empty, with the exception that there was a young man in it who wore a long white dress and told the women to tell the Apostles to go to Galilee, where they should see Christ. On the strength of this we are asked to believe that the reappearance of Christ alive, after a hurried crucifixion, must have been due to supernatural causes, and supernatural causes only! It will be easily seen what a number of threads might be taken up at this point, and followed with not uninteresting results. For the sake, however, of brevity, we grant it as most probable that St. Mark meant the young man said to have been seen in the tomb, to be considered as an angel; but we must also express our conviction that this supposed angelic vision is a misplaced offshoot of the report that Mary Magdalene had seen angels in the tomb after Peter and John had left it.

"It is possible that Mark's account may be the most historic of all those that we have; but we incline to think otherwise, inasmuch as the angelic vision placed in the foreground by Mark and Luke, would not be likely to find its way into the background again, as it does in the fourth Gospel, unless in consequence of really authentic information; no unnecessary detraction from the miraculous element is conceivable as coming from the writer who has handed down to us the story of the raising of Lazarus, where we have, indeed, a real account of a resurrection, the continuity of the evidence being unbroken, and every link in the chain forged fast and strong, even to the unwrapping of the grave clothes from the body as it emerged from the sepulchre. Is it possible that the writer may have given the story of the raising of Lazarus (of which we find no trace except in the fourth Gospel), because he felt that in giving the Apostolic version with absolute or substantial accuracy, he was so weakening the miraculous element in connection with the Resurrection of Jesus Christ himself, that it became necessary to introduce an incontrovertible account of the resurrection of some other person, which should do, as it were, vicarious duty?

"Nevertheless there are some points on which all the three writers are agreed: we have the same substratum of facts, namely, the tomb found already empty when the women reached it, a confused and contradictory report of an angel or angels seen within it, and the subsequent reappearance of Christ. Not one of the three writers affords us the slightest clue as to the time and manner of the removal of the body from the tomb; there is nothing in any of the narratives which is incompatible with its having been taken away on the very night of the Crucifixion itself.

"Is this a case in which the defenders of Christianity would clamour for all the facts, unless they exceedingly well knew that there was no chance of their getting them? All the facts, indeed—what tricks does our imagination play us! One would have thought that there were quite enough facts given as the matter stands to make the defenders of Christianity wish that there were not so many; and then for them to say that if we had more, those that we have would become less contradictory! What right have they to assume that if they had all the facts, the accounts of the Resurrection would cease to puzzle us, more than we have to say that if we had all the facts, we should find these accounts even more inexplicable than we do at present? Had we argued thus we should have been accused of shameless impudence; of a desire to maintain any position in which we happened to find ourselves, and by which we made money, regardless of every common principle of truth or honour, or whatever else makes the difference between upright men and self-deceivers.

"It may be said by some that the discrepancies between the three accounts given above are discrepancies concerning details only, but that all three writers agree about the 'main fact.' We are continually hearing about this 'main fact,' but nobody is good enough to tell us precisely what fact is meant. Is the main fact the fact that Jesus Christ was crucified? Then no one denies it. We all admit that Jesus Christ was crucified. Or, is it that he was seen alive several times after the Crucifixion? This also we are not disposed to deny. We believe that there is a considerable preponderance of evidence in its favour. But if the 'main fact' turns out to be that Christ was crucified, died, and then came to life again, we admit that here too all the writers are agreed, but we cannot find with any certainty that one of them was present when Christ died or when his body was taken down from the Cross, or that there was any such examination of the body as would be absolutely necessary in order to prove that a man had been dead who was afterwards seen alive. If Christ reappeared alive, there is not only no tittle of evidence in support of his death which would be allowed for a moment in an English court of justice, but there is an overwhelming amount of evidence which points inexorably in the direction of his never having died. If he reappeared, there is no evidence of his having died. If he did not reappear, there is no evidence of his having risen from the dead.

"We are inclined, however, as has been said already, to believe that Jesus Christ really did reappear shortly after the Crucifixion, and that his reappearance, though due to natural causes, was conceived to be miraculous. We believe also that Mary fancied that she had seen angels in the tomb, and openly said that she had done so; who would doubt her when so far greater a marvel than this had been made palpably manifest to all? Who would care to inquire very particularly whether there were two angels or only one? Whether there were other women with Mary or whether she was quite alone? Who would compare notes about the exact moment of their appearing, and what strictly accurate account of their words could be expected in the ferment of such excitement and such ignorance? Any speech which sounded tolerably plausible would be accepted under the circumstances, and none will complain of Mark as having wilfully attempted to deceive, any more than he will of Luke: the amplification of the story was inevitable, and the very candour and innocence with which the writers leave loophole after loophole for escape from the miraculous, is alone sufficient proof of their sincerity; nevertheless, it is also proof that they were all more or less inaccurate; we can only say in their defence, that in the reappearance of Christ himself we find abundant palliation of their inaccuracy. Given one great miracle, proved with a sufficiency of evidence for the capacities and proclivities of the age, and the rest is easy. The groundwork of the after-structure of the other miracles is to be found in the fact that Christ was crucified, and was afterwards seen alive."

There is no occasion for me to examine St. Matthew's account of the Resurrection in company with the unhappy men whose views I have been endeavouring to represent above. For reasons which have already been sufficiently dwelt upon I freely own that I agree with them in rejecting it. I shall therefore admit that the story of the sealing of the tomb, and setting of the guard, the earthquake, the descent of the angel from Heaven, his rolling away the stone, sitting upon it, and addressing the women therefrom, is to be treated for all controversial purposes as though it had never been written. By this admission, I confess to complete ignorance of the time when the stone was removed from the mouth of the tomb, or the hour when the Redeemer rose. I should add that I agree with our opponents in believing that our

Lord never foretold His Resurrection to the Apostles. But how little does it matter whether He foretold His Resurrection or not, and whether He rose at one hour or another. It is enough for me that he rose at all; for the rest I care not.

"Yet, see," our opponents will exclaim in answer, "what a mighty river has come from a little spring. We heard first of two men going into an empty tomb, finding two bundles of grave clothes, and departing. Then there comes a certain person, concerning whom we are elsewhere told a fact which leaves us with a very uncomfortable impression, and she sees, not two bundles of grave clothes, but two white angels, who ask a dreamy pointless question, and receive an appropriate answer. Then we find the time of this apparition shifted; it is placed in the front, not in the background, and is seen by many, instead of being vouchsafed to no one but to a weeping woman looking into the bottom of a tomb. The speech of the angels, also, becomes effective, and the linen clothes drop out of sight entirely, unless some faint trace of them is to be found in the 'long white garment' which Mark tells us was worn by the young man who was in the tomb when the women reached it. Finally, we have a guard set upon the tomb, and the stone which was rolled in front of it is sealed; the angel is seen to descend from Heaven, to roll away the stone, and sit upon it, and there is a great earthquake. Oh! how things grow, how things grow! And, oh! how people believe!

"See by what easy stages the story has grown up from the smallest seed, as the mustard tree in the parable, and how the account given by Matthew changes the whole complexion of the events. And see how this account has been dwelt upon to the exclusion of the others by the great painters and sculptors from whom, consciously or unconsciously, our ideas of the Christian era are chiefly drawn. Yes. These men have been the most potent of theologians, for their theology has reached and touched most widely. We have mistaken their echo of the sound for the sound itself, and what was to them an aspiration, has, alas! been to us in the place of science and reality.

"Truly the ease with which the plainest inferences from the Gospel narratives have been overlooked is the best apology for those who have attributed unnatural blindness to the Apostles. If we are so blind, why not they also? A pertinent question, but one which raises more difficulties than it solves. The seeing of truth is as the finding of gold in far countries, where the shepherd has drunk of the stream and used it daily to cleanse the sweat of his brow, and recked little of the treasure which lay abundantly concealed therein, until one luckier than his fellows espies it, and the world comes flocking thither. So with truth; a little care, a little patience, a little sympathy, and the wonder is that it should have lain hidden even from the merest child, not that it should now be manifest.

"How early must it have been objected that there was no evidence that the tomb had not been tampered with (not by the Apostles, for they were scattered, and of him who laid the body in the tomb—Joseph of Arimathæa—we hear no more) and that the body had been delivered not to enemies, but friends; how natural that so desirable an addition to the completeness of the evidences in favour of a miraculous Resurrection should have been early and eagerly accepted. Would not twenty years of oral communication and Spanish or Italian excitability suffice for the rooting of such a story? Yet, as far as we can gather, the Gospel according to St. Matthew was even then unwritten. And who was Matthew? And what was his original Gospel?

"There is one part of his story, and one only, which will stand the test of criticism, and that is this:—That the saying that the disciples came by night and stole the body of Jesus away was current among the Jews, at the time when the Gospel which we now have appeared. Not that they did so—no one will believe this; but the allegation of the rumour (which would hardly have been ventured unless it would command assent as true) points in the direction of search having been made for the body of Jesus—and made in vain.

"We have now seen that there is no evidence worth the name, for any miracle in connection with the tomb of Christ. He probably reappeared alive, but not with any circumstances which we are justified in regarding as supernatural. We are therefore at length led to a consideration of the Crucifixion itself. Is there evidence for more than this—that Christ was crucified, was afterwards seen alive, and that this was regarded by his first followers as a sufficient proof of his having risen from the dead? This would account for the rise of Christianity, and for all the other miracles. Take the following passage from Gibbon:—'The grave and learned Augustine, whose understanding scarcely admits the excuse of credulity, has attested the innumerable prodigies which were worked in Africa by the relics of St. Stephen, and this marvellous narrative is inserted in the elaborate work of "The City of God," which the Bishop designed as a solid and immortal proof of the truth of Christianity. Augustine solemnly declares that he had selected those miracles only which had been publicly certified by persons who were either the objects or the spectators of the powers of the martyr. Many prodigies were omitted or forgotten, and Hippo had been less favourably treated than the other cities of the province, yet the Bishop enumerates above seventy miracles, of which three were resurrections from the dead, within the limits of his own diocese. If we enlarge our view to all the dioceses and all the saints of the Christian world, it will not be easy to calculate the fables and errors which issued from this inexhaustible source. But we may surely be allowed to observe that a miracle in that age of superstition and credulity lost its name and its merits, since it could hardly be considered as a deviation from the established laws of Nature.'— (Gibbon's Decline and Fall, chap. xxviii., sec. 2).

"Who believes in the miracles, or who would dare to quote them? Yet on what better foundation do those of the New Testament rest? For the death of Christ there is no evidence at all. There is evidence that he was believed to have been dead (under circumstances where a misapprehension was singularly likely to arise), by men whose minds were altogether in a different clef to ours as regards the miraculous, and whom we cannot therefore fairly judge by any modern standard. We cannot judge them, but we are bound to weigh the facts which they relate, not in their balance, but in our own. It is not what might have seemed reasonably believable to them, but what is reasonably believable in our own more enlightened age which can be alone accepted sinlessly by ourselves. Men's modes of thought concerning facts change from age to age; but the facts change not at all, and it is of them that we are called to judge.

"We turn to the fourth Gospel, as that from which we shall derive the most accurate knowledge of the facts connected with the Crucifixion. Here we find that it was about twelve o'clock when Pilate brought out Christ for the last time; the dialogue that followed, the preparations for the Crucifixion, and the leading Christ outside the city to the place where the Crucifixion was to take place, could hardly have occupied less than an hour. By six o'clock (by consent of all writers) the body was entombed, so that the actual time during which Christ hung upon the cross was little more than four hours. Let us be thankful to hope that the time of suffering may have been so short—but say five hours, say six, say whatever the reader chooses, the Crucifixion was avowedly too hurried for death in an ordinary case to have ensued. The thieves had to be killed, as yet alive. Immediately before being taken down from the cross the body was delivered to friends. Within thirty-six hours afterwards the tomb in which it had been laid was discovered to have been opened; for how long it had been open we do not know, but a few hours later Christ was seen alive.

"Let it be remembered also that the fact of the body having been delivered to Joseph before the taking down from the cross, greatly enhanced the chance of an escape from death, inasmuch as the duties of the soldiers would have ended with the presentation of the order from Pilate. If any faint symptom of returning animation shewed itself in consequence of the mere change of position and the inevitable shock attendant upon being moved, the soldiers would not know it; their task was ended, and they would not be likely either to wish, or to be allowed, to have anything to do with the matter. Joseph appears to have been a rich man, and would be followed by attendants. Moreover, although we are told by Mark that Pilate sent for the centurion to inquire whether Christ was dead, yet the same writer also tells us that this centurion had already come to the conclusion that Christ was the Son of God, a statement which is supported by the accounts of Matthew and Luke; Mark is the only Evangelist who tells us that the centurion was sent for, but even granting that this was so, would not one who had already recognised Christ as the Son of God be inclined to give him every assistance in his power? He would be frightened, and anxious to get the body down from the cross as fast as possible. So long as Christ appeared to be dead, there would be no unnecessary obstacle thrown in the way of the delivery of the body to Joseph, by a centurion who believed that he had been helping to crucify the Son of God. Besides Joseph was rich, and rich people have many ways of getting their wishes attended to.

"We know of no one as assisting at the taking down or the removal of the body, except Joseph of Arimathæa, for the presence of Nicodemus, and indeed his existence, rests upon the slenderest evidence. None of the Apostles appear to have had anything to do with the deposition, nor yet the women who had come from Galilee, who are represented as seeing where the body was laid (and by Luke as seeing how it was laid), but do not seem to have come into close contact with the body.

"Would any modern jury of intelligent men believe under similar circumstances that the death had been actual and complete? Would they not regard—and ought they not to regard—reappearance as constituting ample proof that there had been no death? Most assuredly, unless Christ had had his head cut off, or had been seen to be burnt to ashes. Again, if unexceptionable medical testimony as to the completeness of the death had reached us, there would be no help for it; we should have to admit that something had happened which was at variance with all our experience of the course of nature; or again if his legs had been broken, or his feet pierced, we could say nothing; but what irreparable mischief is done to any vital function of the body by the mere act of crucifixion? The feet were not always, 'nor perhaps generally,' pierced (so Dean Alford tells us, quoting from Justin Martyr), nor is there a particle of evidence to shew that any exception was made in the present instance. A man who is crucified dies from sheer exhaustion, so that it cannot be deemed improbable that he might swoon away, and that every outward appearance of death might precede death by several hours.

"Are we to suppose that a handful of ignorant soldiers should be above error, when we remember that men have been left for dead, been laid out for burial and buried by their best friends—nay, that they have over and over again been pronounced dead by skilled physicians, when the facilities for knowing the truth were far greater, and when a mistake was much less likely to occur, than at the hurried Crucifixion of Jesus Christ? The soldiers would apply no polished mirror to the lips, nor make use of any of those tests which, under the circumstances, would be absolutely necessary before life could be pronounced to be extinct; they would see that the body was lifeless, inanimate, to all outward appearance like the few other dead bodies which they had probably observed closely; with this they would rest contented.

"It is true, they probably believed Christ to be dead at the time they handed over the body to his friends, and if we had heard nothing more of the matter we might assume that they were right; but the reappearance of Christ alive changes the whole complexion of the story. It is not very likely that the Roman soldiers would have been mistaken in believing him to be dead, unless the hurry of the whole affair, and the order from Pilate, had disposed them to carelessness, and to getting the matter done as fast as possible; but it is much less likely that a dead man should come to life again than that a mistake should have been made about his having being dead. The latter is an event which probably happens every week in one part of the world or another; the former has never yet been known.

"It is not probable that a man officially executed should escape death; but that a dead man should escape from it is more improbable still; in addition to the enormous preponderance of probability on the side of Christ's never having died which arises from this consideration alone, we are told many facts which greatly lessen the improbability of his having escaped death, inasmuch as the Crucifixion was hurried, and the body was immediately delivered to friends without the known destruction of any organic function, and while still hanging upon the cross.

"Joseph and Nicodemus (supposing that Nicodemus was indeed a party to the entombment) may be believed to have thought that Christ was dead when they received the body, but they could not refuse him their assistance when they found out their mistake, nor, again, could they forfeit their high position by allowing it to be known that they had restored the life of one who was so obnoxious to the authorities. They would be in a very difficult position, and would take the prudent course of backing out of the matter at the first moment that humanity would allow, of leaving the rest to chance, and of keeping their own counsel. It is noticeable that we never hear of them again; for there were no two people in the world better able to know whether the Resurrection was miraculous or not, and none who would be more deeply interested in favour of the miracle. They had been faithful when the Apostles themselves had failed, and if their faith had been so strong while everything pointed in the direction of the utter collapse of Christianity, what would it be, according to every natural impulse of self-approbation, when so transcendent a miracle as a resurrection had been worked almost upon their own premises, and upon one whose remains they had generously taken under their protection at a time when no others had ventured to shew them respect?

"We should have fancied that Mary would have run to Joseph and Nicodemus, not to the Apostles; that Joseph and Nicodemus would then have sent for the Apostles, or that, to say the least of it, we should have heard of these two persons as having been prominent members of the Church at Jerusalem; but here again the experience of the ordinary course of nature fails us, and we do not find another word or hint concerning them. This may be the result of accident, but if so, it is a very unfortunate accident, and we have already had a great deal too much of unfortunate accidents, and of truths which may be truths, but which are uncommonly like exaggeration. Stories are like people, whom we judge of in no small degree by the dress they wear, the company they keep, and that subtle indefinable something which we call their expression.

"Nevertheless, there arise the questions how far the spear wound recorded by the writer of the fourth Gospel must be regarded, firstly, as an actual occurrence, and, secondly, as having been necessarily fatal, for unless these things are shewn to be indisputable we have seen that the balance of probability lies greatly in favour of Christ's having escaped with life. If, however, it can be proved that it is a matter of certainty both that the wound was actually inflicted, and that death must have inevitably followed, then the death of Christ is proved. The Resurrection becomes supernatural; the Ascension forthwith ceases to be marvellous; the Miraculous Conception, the Temptation in the Wilderness, all the other miracles of Christ and his Apostles, become believable at once upon so signal a failure of human experience; human experience ceases to be a guide at all, inasmuch as it is found to fail on the very point where it has been always considered to be most firmly established—the remorselessness of the grip of death. But before we can consent to part with the firm ground on which we tread, in the confidence of which we live, move, and have our being—the trust in the established experience of countless ages—we must prove the infliction of the wound and its necessarily fatal character beyond all possibility of mistake. We cannot be expected to reject a natural solution of an event however mysterious, and to adopt a supernatural in its place, so long as there is any element of doubt upon the supernatural side.

"The natural solution of the origin of belief in the Resurrection lies very ready to our hands; once admit that Christ was crucified hurriedly, that there is no proof of the destruction of any organic function of the body, that the body itself was immediately delivered to friends, and that thirty-six hours afterwards Christ was seen alive, and it is impossible to understand how any human being can doubt what he ought to think. We must own also that once let Joseph have kept his own counsel (and he had a great stake to lose if he did not keep it), once let the Apostles believe that Christ's restoration to life was miraculous (and under the circumstances they would be sure to think so), and their reason would be so unsettled that in a very short time all the recognised and all the apocryphal miracles of Christ would pass current with them without a shadow of difficulty."

It will be observed that throughout both this and the preceding chapter I have been dealing with those of our opponents who, while admitting the reappearances of our Lord, ascribe them to natural causes only. I consider this position to be only second in importance to the one taken by Strauss, and as perhaps in some respects capable of being supported with an even greater outward appearance of probability. I therefore resolved to combat it, and as a preliminary to this, have taken care that it shall be stated in the clearest and most definite manner possible. But it is plain that those who accept the fact that our Lord reappeared after the Crucifixion differ hardly less widely from Strauss than they do from ourselves; it will therefore be expedient to shew how they maintain their ground against so formidable an antagonist. Let it be remembered that Strauss and his followers admit that the Death of our Lord is proved, while those of our opponents who would deny this, nevertheless admit that we can establish the reappearances; it follows therefore that each of our most important propositions is admitted by one section or other of the enemy, and each section would probably be heartily glad to be able to deny what it admits. Can there be any doubt about the significance of this fact? Would not a little reflection be likely to suggest to the distracted host of our adversaries that each of its two halves is right, as far as it goes, but that agreement will only be possible between them when each party has learnt that it is in possession of only half the truth, and has come to admit both the Death of our Lord and His Resurrection?

Returning, however, to the manner in which the section of our opponents with whom I am now dealing meet Strauss, they may be supposed to speak as follows:—

"Strauss believes that Christ died, and says (New Life of Jesus, Vol. I., p. 411) that 'the account of the Evangelists of the death of Jesus is clear, unanimous, and connected.' If this means that the Evangelists would certainly know whether Christ died or not, we demur to it at once. Strauss would himself admit that not one of the writers who have recorded the facts connected with the Crucifixion was an eyewitness of that event, and he must also be aware that the very utmost which any of these writers can have known, was that Christ was believed to have been dead. It is strange to see Strauss so suddenly struck with the clearness, unanimity, and connectedness of the Evangelists. In the very next sentence he goes on to say, 'Equally fragmentary, full of contradiction and obscurity, is all that they tell us of the opportunities of observing him which his adherents are supposed to have had after his resurrection.'

Now, this seems very unfair, for, after all, the gospel writers are quite as unanimous in asserting the main fact that Christ reappeared, as they are in asserting that he died; they would seem to be just as 'clear, unanimous, and connected,' about the former event as the latter (for the accounts of the Crucifixion vary not a little), and they must have had infinitely better means of knowing whether Christ reappeared than whether he had actually died. There is not the same scope for variation in the bare assertion that a man died, as there is in the narration of his sayings and doings upon the several occasions of his reappearance. Besides, in support of the reappearances, we have the evidence of Paul, who, though not an eye-witness, was well acquainted with those who were; whereas no man can make more out of the facts recorded concerning the death of Jesus, than that he was believed to be dead under circumstances in which mistake might easily arise, that there is no reason to think that any organic function of the body had been destroyed at the time that it was delivered over to friends, and that none of those who testified to Christ's death appear to have verified their statement by personal inspection of the body. On these points the Evangelists do indeed appear to be 'clear, unanimous, and connected.'

"Later on Strauss is even more unsatisfactory, for on the page which follows the one above quoted from, he writes: 'Besides which, it is quite evident that this (the natural) view of the resurrection of Jesus, apart from the difficulties in which it is involved, does not even solve the problem which is here under consideration: the origin, that is, of the Christian Church by faith in the miraculous resurrection of the Messiah. It is impossible that a being who had stolen half-dead out of a sepulchre, who crept about weak and ill, wanting medical treatment, who required bandaging, strengthening, and indulgence, and who still, at last, yielded to his sufferings, could have given to the disciples the impression that he was a conqueror over death and the grave, the Prince of Life, an impression which lay at the bottom of their future ministry. Such a resuscitation could only have weakened the impression which he had made upon them in life and in death; at the most could only have given it an elegiac voice, but could by no possibility have changed their sorrow into enthusiasm, have elevated their reverence into worship.'

"Now, the fallacy in the above is obvious; it assumes that Christ was in such a state as to be compelled to creep about, weak and ill, &c., and ultimately to die from the effects of his sufferings; whereas there is not a word of evidence in support of all this. He may have been weak and ill when he forbade Mary to touch him, on the first occasion of his being seen alive; but it would be hard to prove even this, and on no subsequent occasion does he shew any sign of weakness. The supposition that he died of the effects of his sufferings is quite gratuitous; one would like to know where Strauss got it from. He may have done so, or he may have been assassinated by some one commissioned by the Jewish Sanhedrim, or he may have felt that his work was done, and that any further interference upon his part would only mar it, and therefore resolved upon withdrawing himself from Palestine for ever, or Joseph of Arimathæa may have feared the revolution which he saw approaching—or twenty things besides might account for Christ's final disappearance. The only thing, however, which we can say with any certainty is that he disappeared, and that there is no reason to believe that he died of his wounds. All over and above this is guesswork.

"Again, if Christ on reappearing had continued in daily intercourse with his disciples, it might have been impossible that they should not find out that he was in all respects like themselves. But he seems to have been careful to avoid seeing them much. Paul only mentions five reappearances, only one of which was to any considerable number of people. According also to the gospel writers, the reappearances were few; they were without preparation, and nothing seems to have been known of where he resided between each visit; this rarity and mysteriousness of the reappearances of Christ (whether dictated by fear of his enemies or by policy) would heighten their effect, and prevent the Apostles from knowing much more about their master than the simple fact that he was indisputably alive. They saw enough to assure them of this, but they did not see enough to prevent their being able to regard their master as a conqueror over death and the grave, even though it could be shewn (which certainly cannot be done) that he continued in infirm health, and ultimately died of his wounds.

"If the Apostles had been highly educated English or German Professors, it might be hard to believe them capable of making any mistake; but they were nothing of the kind; they were ignorant Eastern peasants, living in the very thick of every conceivable kind of delusive influence. Strauss himself supposes their minds to have been so weak and unhinged that they became easy victims to hallucination. But if this was the case, they would be liable to other kinds of credulity, and it seems strange that one who would bring them down so low, should be here so suddenly jealous for their intelligence. There is no reason to suppose that Christ was weak and ill after the first day or two, any more than there is for believing that he died of his wounds. This being so, is it not more simple and natural to believe that the Apostles were really misled by a solid substratum of strange events—a substratum which seems to be supported by all the evidence which we can get—than that the whole story of the appearances of Christ after the Crucifixion should be due to baseless dreams and fancies? At any rate, if the Apostles could be misled by hallucination, much more might they be misled by a natural reappearance, which looked not unlike a supernatural one.

"The belief in the miraculous character of the Resurrection is the central point of the whole Christian system. Let this be once believed, and considering the times, which, it must always be remembered, were in respect of credulity widely different from our own, considering the previous hopes and expectations of the Apostles, considering their education, Oriental modes of thought and speech, familiarity with the ideas of miracle and demonology, and unfamiliarity with the ideas of accuracy and science, and considering also the unquestionable beauty and wisdom of much which is recorded as having been taught by Christ, and the really remarkable circumstances of the case—we say, once let the Resurrection be believed to be miraculous, and the rest is clear; there is no further mystery about the origin of the Christian religion.

"So the matter has now come to this pass, that we are to jeopardise our faith in all human experience, if we are unable to see our way clearly out of a few words about a spear wound, recorded as having been inflicted in a distant country nearly two thousand years ago, by a writer concerning whom we are entirely ignorant, and whose connection with any eye-witness of the events which he records is a matter of pure conjecture. We will see about this hereafter; all that is necessary now is to make sure that we do not jeopardise it, if we do see a way of escape, and this assuredly exists."

I will not pain either the reader or myself by a recapitulation of the arguments which have led our opponents as well as the Dean of Canterbury, and I may add, with due apology, myself, to conclude that nothing is known as to the severity or purpose of the spear wound. The case, therefore, of our adversaries will rest thus:—that there is not only no sufficient reason for believing that Christ died upon the cross, but that there are the strongest conceivable reasons for believing that He did not die; that the shortness of time during which He remained upon the cross, the immediate delivery of the body to friends, and, above all, the subsequent reappearance alive, are ample grounds for arriving at such a conclusion. They add further that it would seem a monstrous supposition to believe that a good and merciful God should have designed to redeem the world by the infliction of such awful misery upon His own Son, and yet determined to condemn every one who did not believe in this design, in spite of such a deficiency of evidence that disbelief would appear to be a moral obligation. No good God, they say, would have left a matter of such unutterable importance in a state of such miserable uncertainty, when the addition of a very small amount of testimony would have been sufficient to establish it.

In the two following chapters I shall show the futility and irrelevancy of the above reasoning—if, indeed, that can be called reasoning which is from first to last essentially unreasonable. Plausible as, in parts, it may have appeared, I have little doubt that the reader will have already detected the greater number of the fallacies which underlie it. But before I can allow myself to enter upon the welcome task of refutation, a few more words from our opponents will yet be necessary. However strongly I disapprove of their views, I trust they will admit that I have throughout expressed them as one who thoroughly understands them. I am convinced that the course I have taken is the only one which can lead to their being brought into the way of truth, and I mean to persevere in it until I have explained the views which they take concerning our Lord's Ascension, with no less clearness than I shewed forth their opinions concerning the Resurrection.

"In St. Matthew's Gospel," they will say, "we find no trace whatever of any story concerning the Ascension. The writer had either never heard anything about the matter at all, or did not consider it of sufficient importance to deserve notice.

"Dean Alford, indeed, maintains otherwise. In his notes on the words, 'And lo! I am with you always unto the end of the world,' he says, 'These words imply and set forth the Ascension'; it is true that he adds, 'the manner of which is not related by the Evangelist': but how do the words quoted, 'imply and set forth' the Ascension? They imply a belief that Christ's spirit would be present with his disciples to the end of time; but how do they set forth the fact that his body was seen by a number of people to rise into the air and actually to mount up far into the region of the clouds?

"The fact is simply this—and nobody can know it better than Dean Alford—that Matthew tells us nothing about the Ascension.

"The last verses of Mark's Gospel are admitted by Dean Alford himself to be not genuine, but even in these the subject is dismissed in a single verse, and although it is stated that Christ was received into Heaven, there is not a single word to imply that any one was supposed to have seen him actually on his way thither.

"The author of the fourth Gospel is also silent concerning the Ascension. There is not a word, nor hint, nor faintest trace of any knowledge of the fact, unless an allusion be detected in the words, 'What and if ye shall see the Son of Man ascending where he was before?' (John vi., 62) in reference to which passage Dean Alford, in his note on Luke xxiv., 52, writes as follows:—'And might not we have concluded from the wording of John vi., 62, that our Lord must have intended an ascension insight of some of those to whom he spoke, and that the Evangelist gives that hint, by recording those words without comment, that he had seen it?' That is to say, we are to conclude that the writer of the fourth Gospel actually saw the Ascension, because he tells us that Christ uttered the words, 'What and if ye shall see the Son of Man ascending where he was before?'

"But who was the author of the fourth Gospel? And what reason is there for thinking that that work is genuine? Let us make another extract from Dean Alford. In his prolegomena, chapter v., section 6, on the genuineness of the fourth Gospel, he writes:—'Neither Papias, who carefully sought out all that Apostles and Apostolic men had related regarding the life of Christ; nor Polycarp, who was himself a disciple of the Apostle John; nor Barnabas, nor Clement of Rome, in their epistles; nor, lastly, Ignatius (in his genuine writings), makes any mention of, or allusion to, this gospel. So that in the most ancient circle of ecclesiastical testimony, it appears to be unknown or not recognised.' We may add that there is no trace of its existence before the latter half of the second century, and that the internal evidence against its genuineness appears to be more and more conclusive the more it is examined.

"St. Paul, when enumerating the last appearances of his master, in a passage where the absence of any allusion to the Ascension is almost conclusive as to his never having heard a word about it, is also silent. In no part of his genuine writings does he give any sign of his having been aware that any story was in existence as to the manner in which Christ was received into Heaven.

"Where, then, does the story come from, if neither Matthew, Mark, John, nor Paul appear to have heard of it?

"It comes from a single verse in St. Luke's Gospel—written more than half a century after the supposed event, when few, or more probably none, of those who were supposed to have seen it were either living or within reach to contradict it. Luke writes (xxiv., 51), 'And it came to pass that while he blessed them, he was parted from them, and carried up into Heaven.' This is the only account of the Ascension given in any part of the Gospels which can be considered genuine. It gives Bethany as the place of the miracle, whereas, if Dean Alford is right in saying that the words of Matthew 'set forth' the Ascension, they set it forth as having taken place on a mountain in Galilee. But here, as elsewhere, all is haze and contradiction. Perhaps some Christian writers will maintain that it happened both at Bethany and in Galilee.

"In his subsequent work, written some sixty or seventy years after the Ascension, St. Luke gives us that more detailed account which is commonly present to the imagination of all men (thanks to the Italian painters), when the Ascension is alluded to. The details, it would seem, came to his knowledge after he had written his Gospel, and many a long year after Matthew and Mark and Paul had written. How he came by the additional details we do not know. Nobody seems to care to know. He must have had them revealed to him, or been told them by some one, and that some one, whoever he was, doubtless knew what he was saying, and all Europe at one time believed the story, and this is sufficient proof that mistake was impossible.

"It is indisputable that from the very earliest ages of the Church there existed a belief that Christ was at the right hand of God; but no one who professes to have seen him on his way thither has left a single word of record. It is easy to believe that the facts may have been revealed in a night vision, or communicated in one or other of the many ways in which extraordinary circumstances are communicated, during the years of oral communication and enthusiasm which elapsed between the supposed Ascension of Christ and the writing of Luke's second work. It is not surprising that a firm belief in Christ's having survived death should have arisen in consequence of the actual circumstances connected with the Crucifixion and entombment. Was it then strange that this should develop itself into the belief that he was now in Heaven, sitting at the right hand of God the Father? And finally was it strange that a circumstantial account of the manner in which he left this earth should be eagerly accepted?"

CHAPTER IX

I have completed a task painful to myself and the reader. Painful to myself inasmuch as I am humiliated upon remembering the power which arguments, so shallow and so easily to be refuted, once had upon me; painful to the reader, as everything must be painful which even appears to throw doubt upon the most sublime event that has happened in human history. How little does all that has been written above touch the real question at issue, yet, what self-discipline and mental training is required before we learn to distinguish the essential from the unessential.

Before, however, we come to close quarters with our opponents concerning the views put forward in the preceding chapters, it will be well to consider two questions of the gravest and most interesting character, questions which will probably have already occurred to the reader with such force as to demand immediate answer. They are these.

Firstly, what will be the consequences of admitting any considerable deviation from historical accuracy on the part of the sacred writers?

Secondly, how can it be conceivable that God should have permitted inaccuracy or obscurity in the evidence concerning the Divine commission of His Son?

If God so loved the World that He sent His only begotten Son into it to rescue those who believed in Him from destruction, how is it credible that He should not have so arranged matters as that all should find it easy to believe? If He wanted to save mankind and knew that the only way in which mankind could be saved was by believing certain facts, how can it be that the records of the facts should have been allowed to fall into confusion?

To both these questions I trust that the following answers may appear conclusive.

I. As regards the consequences which may be supposed to follow upon giving up any part of the sacred writings, no matter how seemingly unimportant, it is undoubtedly true that to many minds they have appeared too dangerous to be even contemplated. Thus through fear of some supposed unutterable consequences which would happen to the cause of truth if truth were spoken, people profess to believe in the genuineness of many passages in the Bible which are universally acknowledged by competent judges of every shade of theological opinion to be interpolations into the original text. To say nothing of the Old Testament, where many whole books are of disputed genuineness or authenticity, there are portions of the New which none will seriously defend;—for example, the last verses of St. Mark's Gospel,—containing, as they do, the sentence of damnation against all who do not believe—the second half of the third, and the whole of the fourth verse of the fifth chapter of St. John's Gospel, the story of the woman taken in adultery, and probably the whole of the last chapter of St. John's Gospel, not to mention the Epistle to the Hebrews, the Epistles to Timothy, Titus, and to the Ephesians, the Epistles of Peter and James, the famous verses as to the three witnesses in the First Epistle of St. John, and perhaps also the book of Revelation. These are passages and works about which there is either no doubt at all as to their not being genuine, or over which there hangs so much uncertainty that no dependence can be placed upon them.

But over and above these, there are not a few parts of each of the Gospels which, though of undisputed genuineness, cannot be accepted as historical; thus the account of the Resurrection given by St. Matthew, and parts of those by Luke and Mark, the cursing of the barren fig-tree, and the prophecies of

His Resurrection ascribed to our Lord Himself, will not stand the tests of criticism which we are bound to apply to them if we are to exercise the right of private judgement; instead of handing ourselves over to a priesthood as the sole custodians and interpreters of the Bible. It has been said by some that the miracle of the penny found in the fish's mouth should be included in the above category, but it should be remembered that we have only the injunction of our Lord to St. Peter that he should catch the fish and the promise that he should find the penny in its mouth, but that we have no account of the sequel, it is therefore possible that in the event of St. Peter's faith having failed him he may have procured the money from some other source, and that thus the miracle, though undoubtedly intended, was never actually performed. How unnecessary therefore as well as presumptuous are the Rationalistic interpretations which have been put upon the event by certain German writers!

Now there are few, if any, who would be so illiberal as to wish for the exclusion from the sacred volume of all those books or passages which, though neither genuine nor perhaps edifying, have remained in the Canon of Scripture for many centuries. Any serious attempt to reconstruct the Canon would raise a theological storm which would not subside in this century. The work could never be done perfectly, and even if it could, it would have to be done at the expense of tearing all Christendom in pieces. The passages do little or no harm where they are, and have received the sanction of time; let them therefore by all means remain in their present position. But the question is still forced upon us whether the consequences of openly admitting the certain spuriousness of many passages, and the questionable nature of others as regards morality, genuineness and authenticity, should be feared as being likely to prejudice the main doctrines of Christianity.

The answer is very plain. He who has vouchsafed to us the Christian dispensation may be safely trusted to provide that no harm shall happen, either to it or to us, from an honest endeavour to attain the truth concerning it. What have we to do with consequences? These are in the hands of God. Our duty is to seek out the truth in prayer and humility, and when we believe that we have found it, to cleave to it through evil and good report; to fail in this is to fail in faith; to fail in faith is to be an infidel. Those who suppose that it is wiser to gloss over this or that, and who consider it "injudicious" to announce the whole truth in connection with Christianity, should have learnt by this time that no admission which can by any possibility be required of them can be so perilous to the cause of Christ as the appearance of shirking investigation. It has already been insisted upon that cowardice is at the root of the infidelity which we see around us; the want of faith in the power of truth which exists in certain pious but timid hearts has begotten utter unbelief in the minds of all superficial investigators into Christian evidences. Such persons see that the defenders have something in the background, something which they would cling to although they are secretly aware that they cannot justly claim it. This is enough for many, and hence more harm is done by fear than could ever have been done by boldness. Boldness goes out into the fight, and if in the wrong gets slain, childless. Fear stays at home and is prolific of a brood of falsehoods.

It is immoral to regard consequences at all, where truth and justice are concerned; the being impregnated with this conviction to the inmost core of one's heart is an axiom of common honesty—one of the essential features which distinguish a good man from a bad one. Nevertheless, to make it plain that the consequences of outspoken truthfulness in connection with the scriptural writings would have no harmful effect whatever, but would, on the contrary, be of the utmost service as removing a stumbling-block from the way of many—let us for the moment suppose that very much more would have to be given up than can ever be demanded.

Suppose we were driven to admit that nothing in the life of our Lord can be certainly depended upon beyond the facts that He was begotten by the Holy Ghost of the Virgin Mary; that He worked many miracles upon earth, and delivered St. Matthew's version of the sermon on the mount and most of the parables as we now have them; finally, that He was crucified, dead, and buried, that He rose again from the dead upon the third day, and ascended unto Heaven. Granting for the sake of argument that we could rely on no other facts, what would follow? Nothing which could in any way impair the living power of Christianity.

The essentials of Christianity, i.e., a belief in the Divinity of the Saviour and in His Resurrection and Ascension, have stood, and will stand, for ever against any attacks that can be made upon them, and these are probably the only facts in which belief has ever been absolutely necessary for salvation; the answer, therefore, to the question what ill consequences would arise from the open avowal of things which every student must know to be the fact concerning the biblical writings is that there would be none at all. The Christ-ideal which, after all, is the soul and spirit of Christianity would remain precisely where it was, while its recognition would be far more general, owing to the departure on the part of its apologists from certain lines of defence which are irreconcilable with the ideal itself.

II. Returning to the objection how it could be possible that God should have left the records of our Lord's history in such a vague and fragmentary condition, if it were really of such intense importance for the world to understand it and believe in it, we find ourselves face to face with a question of far greater importance and difficulty.

The old theory that God desired to test our faith, and that there would be no merit in believing if the evidence were such as to commend itself at once to our understanding, is one which need only be stated to be set aside. It is blasphemy against the goodness of God to suppose that He has thus laid as it were an ambuscade for man, and will only let him escape on condition of his consenting to violate one of the very most precious of God's own gifts. There is an ingenious cruelty about such conduct which it is revolting even to imagine. Indeed, the whole theory reduces our Heavenly Father to a level of wisdom and goodness far below our own; and this is sufficient answer to it.

But when, turning aside from the above, we try to adopt some other and more reasonable view, we naturally set ourselves to consider why the Almighty should have required belief in the Divinity of His Son from man. What is there in this belief on man's part which can be so grateful to God that He should make it a sine quâ non for man's salvation? As regards Himself, how can it matter to Him what man should think of Him? Nay, it must be for man's own good that the belief is demanded.

And why? Surely we can see plainly that it is the beauty of the Christ-ideal which constitutes the working power of Christianity over the hearts and lives of men, leading them to that highest of all worships which consists in imitation. Now the sanction which is given to this ideal by belief in the Divinity of our Lord, raises it at once above all possibility of criticism. If it had not been so sanctioned it might have been considered open to improvement; one critic would have had this, and another that; comparison would have been made with ideals of purely human origin such as the Greek ideal, exemplified in the work of Phidias, and in later times with the mediæval Italian ideal, as deducible from the best fifteenth and early sixteenth Italian painting and sculpture, the Madonnas of Bellini and Raphael, or the St. George of Donatello; or again with the ideal derivable from the works of our own Shakespeare, and there are some even now among those who deny the Divinity of Christ who will profess that each one of these ideals is more universal, more fitted for the spiritual food of a man, and indeed actually higher, than that presented by the life and death of our Saviour. But once let the Divine origin of this last ideal be admitted, and there can be no further uncertainty; hence the absolute necessity for belief in Christ's Divinity as closing the most important of all questions, Whereunto should a man endeavour to liken both himself and his children?

Seeing then that we have reasonable ground for thinking that belief in the Divinity of our Lord is mainly required of us in order to exalt our sense of the paramount importance of following and obeying the life and commands of Christ, it is natural also to suppose that whatever may have happened to the records of that life should have been ordained with a view to the enhancing of the preciousness of the ideal.

Now, the fragmentary character, and the partial obscurity—I might have almost written, the incomparable chiaroscuro—of the Evangelistic writings have added to the value of our Lord's character as an ideal, not only in the case of Christians, but as bringing the Christ-ideal within the reach and comprehension of an infinitely greater number of minds than it could ever otherwise have appealed to. It is true that those who are insensible to spiritual influences, and whose materialistic instinct leads them to deny everything which is not as clearly demonstrable by external evidence as a fact in chemistry, geography, or mathematics, will fail to find the hardness, definition, tightness, and, let me add, littleness of outline, in which their souls delight; they will find rather the gloom and gleam of Rembrandt, or the golden twilight of the Venetians, the losing and the finding, and the infinite liberty of shadow; and this they hate, inasmuch as it taxes their imagination, which is no less deficient than their power of sympathy; they would have all found, as in one of those laboured pictures wherein each form is as an inflated bladder and, has its own uncompromising outline remorselessly insisted upon.

Looking to the ideals of purely human creation which have come down to us from old times, do we find that the Theseus suffers because we are unable to realise to ourselves the precise features of the original? Or again do the works of John Bellini suffer because the hand of the painter was less dexterous than his intention pure? It is not what a man has actually put upon his canvas, but what he makes us feel that he felt, which makes the difference between good and bad in painting. Bellini's hand was cunning enough to make us feel what he intended, and did his utmost to realise; but he has not realised it, and the same hallowing effect which has been wrought upon the Theseus by decay (to the enlarging of its spiritual influence), has been wrought upon the work of Bellini by incapacity—the incapacity of the painter to utter perfectly the perfect thought which was within. The early Italian paintings have that stamp of individuality upon them which assures us that they are not only portraits, but as faithful portraits as the painter could make them, more than this we know not, but more is unnecessary.

Do we not detect an analogy to this in the records of the Evangelists? Do we not see the child-like unself-seeking work of earnest and loving hearts, whose innocence and simplicity more than atone for their many shortcomings, their distorted renderings, and their omissions? We can see through these things as through a glass darkly, or as one looking upon some ineffable masterpiece of Venetian portraiture by the fading light of an autumnal evening, when the beauty of the picture is enhanced a hundredfold by the gloom and mystery of dusk. We may indeed see less of the actual lineaments themselves, but the echo is ever more spiritually tuneful than the sound, and the echo we find within us. Our imagination is in closer communion with our longings than the hand of any painter.

Those who relish definition, and definition only, are indeed kept away from Christianity by the present condition of the records, but even if the life of our Lord had been so definitely rendered as to find a place in their system, would it have greatly served their souls? And would it not repel hundreds and thousands of others, who find in the suggestiveness of the sketch a completeness of satisfaction, which no photographic reproduction could have given? The above may be difficult to understand, but let me earnestly implore the reader to endeavour to master its import.

People misunderstand the aim and scope of religion. Religion is only intended to guide men in those matters upon which science is silent. God illumines us by science as with a mechanical draughtsman's plan; He illumines us in the Gospels as by the drawing of a great artist. We cannot build a "Great Eastern" from the drawings of the artist, but what poetical feeling, what true spiritual emotion was ever kindled by a mechanical drawing? How cold and dead were science unless supplemented by art and by religion! Not joined with them, for the merest touch of these things impairs scientific value—which depends essentially upon accuracy, and not upon any feeling for the beautiful and lovable. In like manner the merest touch of science chills the warmth of sentiment—the spiritual life. The mechanical drawing is spoiled by being made artistic, and the work of the artist by becoming mechanical. The aim of the one is to teach men how to construct, of the other how to feel.

For the due conservation therefore of both the essential requisites of human well-being—science, and religion—it is requisite that they be kept asunder and reserved for separate use at different times. Religion is the mistress of the arts, and every art which does not serve religion truly is doomed to perish as a lying and unprofitable servant. Science is external to religion, being a separate dispensation, a distinct revelation to mankind, whereby we are put into full present possession of more and more of God's modes of dealing with material things, according as we become more fitted to receive them through the apprehension of those modes which have been already laid open to us.

We ought not therefore to have expected scientific accuracy from the Gospel records—much less should we be required to believe that such accuracy exists. Does any great artist ever dream of aiming directly at imitation? He aims at representation—not at imitation. In order to attain true mastery here, he must spend years in learning how to see; and then no less time in learning how not to see. Finally, he learns how to translate. Take Turner for example. Who conveys so living an impression of the face of nature? Yet go up to his canvas and what does one find thereon? Imitation? Nay—blotches and daubs of paint; the combination of these daubs, each one in itself when taken alone absolutely untrue, forms an impression which is quite truthful. No combination of minute truths in a picture will give so faithful a representation of nature as a wisely arranged tissue of untruths.

Absolute reproduction is impossible even to the photograph. The work of a great artist is far more truthful than any photograph; but not even the greatest artist can convey to our minds the whole truth of nature; no human hand nor pigments can expound all that lies hidden in "Nature's infinite book of secrecy"; the utmost that can be done is to convey an impression, and if the impression is to be conveyed truthfully, the means must often be of the most unforeseen character. The old Pre-Raphaelites aimed at absolute reproduction. They were succeeded by a race of men who saw all that their predecessors had seen, but also something higher. The Van Eycks and Memling paved the way for painters who found their highest representatives in Rubens, Vandyke, and Rembrandt—the mightiest of them all. Giovanni Bellini, Carpaccio and Mantegna were succeeded by Titian, Giorgione, and Tintoretto; Perugino was succeeded by Raphael. It is everywhere the same story; a reverend but child-like worship of the letter, followed by a manful apprehension of the spirit, and, alas! in due time by an almost total disregard of the letter; then rant and cant and bombast, till the value of the letter is reasserted. In theology the early men are represented by the Evangelicals, the times of utter decadence by infidelity—the middle race of giants is yet to come, and will be found in those who, while seeing something far beyond either minute accuracy or minute inaccuracy, are yet fully alive both to the letter and to the spirit of the Gospels.

Again, do not the seeming wrongs which the greatest ideals of purely human origin have suffered at the hands of time, add to their value instead of detracting from it? Is it not probable that if we were to see the glorious fragments from the Parthenon, the Theseus and the Ilyssus, or even the Venus of Milo, in their original and unmutilated condition, we should find that they appealed to us much less forcibly than they do at present? All ideals gain by vagueness and lose by definition, inasmuch as more scope is left for the imagination of the beholder, who can thus fill in the missing detail according to his own spiritual needs. This is how it comes that nothing which is recent, whether animate or inanimate, can serve as an ideal unless it is adorned by more than common mystery and uncertainty. A new Cathedral is necessarily very ugly. There is too much found and too little lost. Much less could an absolutely perfect Being be of the highest value as an ideal, as long as He could be clearly seen, for it is impossible that He could be known as perfect by imperfect men, and His very perfections must perforce appear as blemishes to any but perfect critics. To give therefore an impression of perfection, to create an absolutely unsurpassable ideal, it became essential that the actual image of the original should become blurred and lost, whereon the beholder now supplies from his own imagination that which is to him more perfect than the original, though objectively it must be infinitely less so.

It is probably to this cause that the incredulity of the Apostles during our Lord's life-time must be assigned. The ideal was too near them, and too far above their comprehension; for it must be always remembered that the convincing power of miracles in the days of the Apostles must have been greatly weakened by the current belief in their being events of no very unusual occurrence, and in the existence both of good and evil spirits who could take possession of men and compel them to do their bidding. A resurrection from the dead or a restoration of sight to the blind, must have seemed even less portentous to them, than an unusually skilful treatment of disease by a physician is to us. We can therefore understand how it happened that the faith of the Apostles was so little to be depended upon even up to the Crucifixion, inasmuch as the convincing power of miracles had been already, so to speak, exhausted, a fact which may perhaps explain the early withdrawal of the power to work them; we cannot indeed believe that it could have been so far weakened as to make the Apostles disregard the prophecies of their Master that He should rise from the dead, if He had ever uttered them, and we have already seen reason to think that these prophecies are the ex post facto handiwork of time; but the incredulity of the disciples, when seen through the light now thrown upon it, loses that wholly inexplicable character which it would otherwise bear.

But to return to the subject of the ideal presented by the life and death of our Lord. In the earliest days of the Church there can have been no want of the most complete and irrefragable evidence for the objective reality of the miracles, and especially of the Resurrection and Ascension. The character of Christ would also stand out revealed to all, with the most copious fulness of detail. The limits within which so sharply defined an ideal could be acceptable were narrow, but as the radius of Christian influence increased, so also would the vagueness and elasticity of the ideal; and as the elasticity of the ideal, so also the range of its influence.

A beneficent and truly marvellous provision for the greater complexity of man's spiritual needs was thus provided by a gradual loss of detail and gain of breadth. Enough evidence was given in the first instance to secure authoritative sanction for the ideal. During the first thirty or forty years after the death of our Lord no one could be in want of evidence, and the guilt of unbelief is therefore brought prominently forward. Then came the loss of detail which was necessary in order to secure the universal acceptability of the ideal; but the same causes which blurred the distinctness of the features, involved the inevitable blurring of no small portions of the external evidences whereby the Divine origin of the ideal was established. The primary external evidence became less and less capable of compelling instantaneous assent, according as it was less wanted, owing to the greater mass of secondary evidence, and to the growth of appreciation of the internal evidences, a growth which would be fostered by the growing adaptability of the ideal.

Some thirty or forty years, then, from the death of our Saviour the case would stand thus. The Christ-ideal would have become infinitely more vague, and hence infinitely more universal: but the causes which had thus added to its value would also have destroyed whatever primary evidence was superabundant, and the vagueness which had overspread the ideal would have extended itself in some measure over the evidences which had established its Divine origin.

But there would of course be limits to the gain caused by decay. Time came when there would be danger of too much vagueness in the ideal, and too little distinctness in the evidences. It became necessary therefore to provide against this danger.

Precisely at that epoch the Gospels made their appearance. Not simultaneously, not in concert, and not in perfect harmony with each other, yet with the error distributed skilfully among them, as in a well-tuned instrument wherein each string is purposely something out of tune with every other. Their divergence of aim, and different authorship, secured the necessary breadth of effect when the accounts were viewed together; their universal recognition afforded the necessary permanency, and arrested further decay. If I may be pardoned for using another illustration, I would say that as the roundness of the stereoscopic image can only be attained by the combination of two distinct pictures, neither of them in perfect harmony with the other, so the highest possible conception of Christ, cannot otherwise be produced than through the discrepancies of the Gospels.

From the moment of the appearing of the Gospels, and, I should add, of the Epistles of St. Paul, the external evidences of Christianity became secured from further change; as they were then, so are they now, they can neither be added to nor subtracted from; they have lain as it were sleeping, till the time should come to awaken them. And the time is surely now, for there has arisen a very numerous and increasing class of persons, whose habits of mind unfit them for appreciating the value of vagueness, but who have each one of them a soul which may be lost or saved, and on whose behalf the evidences for the authority whereby the Christ-ideal is sanctioned, should be restored to something like their former sharpness. Christianity contains provision for all needs upon their arising. The work of restoration is easy. It demands this much only—the recognition that time has made incrustations upon some parts of the evidences, and that it has destroyed others; when this is admitted, it becomes easy, after a little practice, to detect the parts that have been added, and to remove them, the parts that are wanting, and to supply them. Only let this be done outside the pages of the Bible itself, and not to the disturbance of their present form and arrangement.

The above explanation of the causes for the obscurity which rests upon much of our Lord's life and teaching, may give us ground for hoping that some of those who have failed to feel the force of the external evidences hitherto, may yet be saved, provided they have fully recognised the Christ-ideal and endeavoured to imitate it, although irrespectively of any belief in its historical character.

It is reasonable to suppose that the duty of belief was so imperatively insisted upon, in order that the ideal might thus be exalted above controversy, and made more sacred in the eyes of men than it could have been if referable to a purely human source. May not, then, one who recognises the ideal as his summum bonum find grace although he knows not, or even cares not, how it should have come to be so? For even a sceptic who regarded the whole New Testament as a work of art, a poem, a pure fiction from beginning to end, and who revered it for its intrinsic beauty only, as though it were a picture or statue, even such a person might well find that it engendered in him an ideal of goodness and power and love and human sympathy, which could be derived from no other source. If, then, our blessed Lord so causes the sun of His righteousness to shine upon these men, shall we presume to say that He will not in another world restore them to that full communion with Himself which can only come from a belief in His Divinity?

We can understand that it should have been impossible to proclaim this in the earliest ages of the Church, inasmuch as no weakening of the sanctions of the ideal could be tolerated, but are we bound to extend the operation of the many passages condemnatory of unbelief to a time so remote as our own, and to circumstances so widely different from those under which they were uttered? Do we so extend the command not to eat things strangled or blood, or the assertion of St. Paul that the unmarried state is higher than the married? May we not therefore hope that certain kinds of unbelief have become less hateful in the sight of God inasmuch as they are less dangerous to the universal acceptance of our Lord as the one model for the imitation of all men? For, after all, it is not belief in the facts which constitutes the essence of Christianity, but rather the being so impregnated with love at the contemplation of Christ that imitation becomes almost instinctive; this it is which draws the hearts of men to God the Father, far more than any intellectual belief that God sent our Lord into the world, ordaining that he should be crucified and rise from the dead. Christianity is addressed rather to the infinite spirit of man than to his finite intelligence, and the believing in Christ through love is more precious in the sight of God than any loving through belief. May we not hope, then, that those whose love is great may in the end find acceptance, though their belief is small? We dare not answer this positively; but we know that there are times of transition in the clearness of the Christian evidences as in all else, and the treatment of those whose lot is cast in such times will surely not escape the consideration of our Heavenly Father.

But with reference to the many-sidedness of the Christ-ideal, as having been part of the design of God, and not attainable otherwise than as the creation of destruction—as coming out of the waste of time— it is clear that the perception of such a design could only be an offspring of modern thought; the conception of such an apparently self-frustrating scheme could only arise in minds which were familiar with the manner in which it is necessary "to hound nature in her wanderings" before her feints can be eluded, and her prevarications brought to book. A deep distrust of the over-obvious is wanted, before men can be brought to turn aside from objections which at the first blush appear to be very serious, and to take refuge in solutions which seem harder than the problems which they are intended to solve. What a shock must the discovery of the rotation of the earth have given to the moral sense of the age in which it was made. How it contradicted all human experience. How it must have outraged common sense. How it must have encouraged scepticism even about the most obvious truths of morality. No question could henceforth be considered settled; everything seemed to require reopening; for if man had once been deceived by Nature so entirely, if he had been so utterly led astray and deluded by the plausibility of her pretence that the earth was immovably fixed, what else, that seemed no less incontrovertible, might not prove no less false?

It is probable that the opposition to Galileo on the part of the Roman church was as much due to some such feelings as these, as to theological objections; the discovery was felt to unsettle not only the foundations of the earth, but those of every branch of human knowledge and polity, and hence to be an outrage upon morality itself. A man has no right to be very much in advance of other people; he is as a sheep, which may lead the mob, but must not stray forward a quarter of a mile in front of it; if he does this, he must be rounded up again, no matter how right may have been his direction. He has no right to be right, unless he can get a certain following to keep him company; the shock to morality and the encouragement to lawlessness do more harm than his discovery can atone for. Let him hold himself back till he can get one or two more to come with him. In like manner, had reflections as to the advantage gained by the Christ ideal in consequence of the inaccuracies and inconsistencies of the Gospels—reflections which must now occur to any one—been put forward a hundred years ago, they would have met justly with the severest condemnation. But now, even those to whom they may not have occurred already will have little difficulty in admitting their force.

But be this as it may, it is certain that the inability to understand how the sense of Christ in the souls of men could be strengthened by the loss of much knowledge of His character, and of the facts connected with His history, lies at the root of the error even of the Apostle St. Paul, who exclaims with his usual fervour, but with less than his usual wisdom, "Has Christ been divided?" (I. Cor. i., 13). "Yea," we may make answer, "He is divided and is yet divisible that all may share in Him." St. Paul himself had realised that it was the spiritual value of the Christ-ideal which was the purifier and refresher of our souls, inasmuch as he elsewhere declares that even though he had known Christ Himself after the flesh, he knew Him no more; the spiritual Christ, that is to say the spirit of Christ as recognisable by the spirits of men, was to him all in all. But he lived too near the days of our Lord for a full comprehension of the Christian scheme, and it is possible that had he known Christ after the flesh, his soul might have been less capable of recognising the spiritual essence, rather than more so. Have we here a faint glimmering of the motive of the Almighty in not having allowed the Gentile Apostle to see Christ after the flesh? We cannot say. But we may say this much with certainty, that had he been living now, St. Paul would have rejoiced at the many-sidedness of Christ, which he appears to have hardly recognised in his own life-time.

The apparently contradictory portraits of our Lord which we find in the Gospels—so long a stumbling-block to unbelievers—are now seen to be the very means which enable men of all ranks, and all shades of opinion, to accept Christ as their ideal; they are like the sea, which from having seemed the most impassable of all objects, turns out to be the greatest highway of communication. To the artisan, for instance, who may have long been out of work, or who may have suffered from the greed and selfishness of his employers, or again, to the farm labourer who has been discharged perhaps at the approach of winter, the parable of "the Labourers in the Vineyard" offers itself as a divinely sanctioned picture of the dealings of God with man; few but those who have mixed much with the less educated classes, can have any idea of the priceless comfort which this parable affords daily to those whose lot it has been to remain unemployed when their more fortunate brethren have been in full work. How many of the poor, again, are drawn to Christianity by the parable of Dives and Lazarus. How many a humble-minded Christian while reflecting upon the hardness of his lot, and tempted to cast a longing eye upon the luxuries which are at the command of his richer neighbours, is restrained from seriously coveting them, by remembering the awful fate of Dives, and the happy future which was in store for Lazarus. "Dives," they exclaim, "in his life-time possessed good things and in like manner Lazarus evil things, but now the one is comforted in the bosom of Abraham, and the other tormented in a lake of fire." They remember, also, that it is easier for a camel to go through the eye of a needle than for a rich man to enter into the kingdom of Heaven.

It has been said by some that the poor are thus encouraged to gloat over the future misery of the rich, and that many of the sayings ascribed to our Lord have an unhealthy influence over their minds. I remember to have thought so once myself, but I have seen reason to change my mind. Hope is given by these sayings to many whose lives would be otherwise very nearly hopeless, and though I fully grant that the parable of Dives and Lazarus can only afford comfort to the very poor, yet it is most certain that it does afford comfort to this numerous class, and helps to keep them contented with many things which they would not otherwise endure.

On the other hand, though the poor are first provided for, the rich are not left without their full share of consolation. Joseph of Arimathæa was rich, and modern criticism forbids us to believe that the parable of Dives and Lazarus was ever actually spoken by our Lord—at any rate not in its present form. Neither are the children of the rich forgotten; the son who repents at length of a course of extravagant or riotous living is encouraged to return to virtue, and to seek reconciliation with his father, by reflecting upon the parable of the Prodigal Son, wherein he will find an everlasting model for the conduct of all earthly fathers. I will say nothing of the parable of the Unjust Steward, for it is one of which the interpretation is most uncertain; nevertheless I am sure that it affords comfort to a very large number of persons.

Christ came not to the whole, but to those that were sick; he came not to call the righteous but sinners to repentance. Even our fallen sisters are remembered in the story of the woman taken in adultery, which reminds them that they can only be condemned justly by those who are without sin. It is to the poor, the weak, the ignorant and the infirm that Christianity appeals most strongly, and to whose needs it is most especially adapted—but these form by far the greater portion of mankind. "Blessed are they that mourn!" Whose sorrow is not assuaged by the mere sound of these words? Who again is not reassured by being reminded that our Heavenly Father feeds the sparrows and clothes the lilies of the field, and that if we will only seek the kingdom of God and His righteousness we need take no heed for the morrow what we shall eat, and what we shall drink, nor wherewithal we shall be clothed. God will provide these things for us if we are true Christians, whether we take heed concerning them or not. "I have been young and now am old," saith the Psalmist, "yet never saw I the righteous forsaken nor his seed begging their bread."

How infinitely nobler and more soul-satisfying is the ideal of the Christian saint with wasted limbs, and clothed in the garb of poverty—his upturned eyes piercing the very heavens in the ecstasy of a divine despair—than any of the fleshly ideals of gross human conception such as have already been alluded to. If a man does not feel this instinctively for himself, let him test it thus—whom does his heart of hearts tell him that his son will be most like God in resembling? The Theseus? The Discobolus? or the St. Peters and St. Pauls of Guido and Domenichino? Who can hesitate for a moment as to which ideal presents the higher development of human nature? And this I take it should suffice; the natural instinct which draws us to the Christ-ideal in preference to all others as soon as it has been once presented to us, is a sufficient guarantee of its being the one most tending to the general well-being of the world.

CHAPTER X

CONCLUSION

It only remains to return to the seventh and eighth chapters, and to pass in review the reasons which will lead us to reject the conclusions therein expressed by our opponents.

These conclusions have no real bearing upon the question at issue. Our opponents can make out a strong case, so long as they confine themselves to maintaining that exaggeration has to a certain extent impaired the historic value of some of the Gospel records of the Resurrection. They have made out this much, but have they made out more? They have mistaken the question—which is this—"Did Jesus Christ die and rise from the dead?" And in the place of it they have raised another, namely, "Has there been any inaccuracy in the records of the time and manner of His reappearing?"

Our error has been that instead of demurring to the relevancy of the issue raised by our opponents, we have accepted it. We have thus placed ourselves in a false position, and have encouraged our opponents by doing so. We have undertaken to fight them upon ground of their own choosing. We have been discomfited; but instead of owning to our defeat, and beginning the battle anew from a fresh base of operations, we have declared that we have not been defeated; hence those lamentable and suicidal attempts at disingenuous reasoning which we have seen reason to condemn so strongly in the works of Dean Alford and others. How deplorable, how unchristian they are!

The moment that we take a truer ground, the conditions of the strife change. The same spirit of candid criticism which led us to reject the account of Matthew in toto, will make it easy for us to admit that those of Mark, Luke, and John, may not be so accurate as we could have wished, and yet to feel that our cause has sustained no injury. There are probably very few who would pin their faith to the fact that Julius Cæsar fell exactly at the feet of Pompey's statue, or that he uttered the words "Et tu, Brute." Yet there are still fewer who would dispute the fact that Julius Caesar was assassinated by conspirators of whom Brutus and Cassius were among the leaders. As long as we can be sure that our Lord died and rose from the dead, we may leave it to our opponents to contend about the details of the manner in which each event took place.

We had thought that these details were known, and so thinking, we had a certain consolation in realising to ourselves the precise manner in which every incident occurred; yet on reflection we must feel that the desire to realise is of the essence of idolatry, which, not content with knowing that there is a God, will be satisfied with nothing if it has not an effigy of His face and figure. If it has not this it falls straight-way to the denial of God's existence, being unable to conceive how a Being should exist and yet be incapable of representation. We are as those who would fall down and worship the idol; our opponents, as those who upon the destruction of the idol would say that there was no God.

We have met sceptics hitherto by adhering to the opinions as to the necessity of accuracy which prevailed among our forefathers, and instead of saying, "You are right—we do not know all that we thought we did—nevertheless we know enough—we know the fact, though the manner of the fact be hidden," we have preferred to say, "You are mistaken, our severe outline, our hard-and-fast lines are all perfectly accurate, there is not a detail of our theories which we are not prepared to stand by." On this comes recrimination and mutual anger, and the strife grows hotter and hotter.

Let us now rather say to the unbeliever, "We do not deny the truth of much which you assert. We give up Matthew's account of the Resurrection; we may perhaps accept parts of those of Mark and Luke and John, but it is impossible to say which parts, unless those in which all three agree with one another; and this being so, it becomes wiser to regard all the accounts as early and precious memorials of the certainty felt by the Apostles that Christ died and rose again, but as having little historic value with regard to the time and manner of the Resurrection."

Once take this ground, and instead of demurring to the truth of many of the assertions of our opponents, demur to their relevancy, and the unbeliever will find the ground cut away from under his feet independently of the fact that the reasonableness of the concession, and the discovery that we are not fighting merely to maintain a position, will incline him to calmness and to the reconsideration of his own opinions—which will in itself be a great gain—he will soon perceive that we are really standing upon firm ground, from which no enemy can dislodge us. The discovery that we know less of the time and manner of our Lord's death and Resurrection than we thought we did, does not invalidate a single one of the irresistible arguments whereby we can establish the fact of His having died and risen again. The reader will now perhaps begin to perceive that the sad division between Christians and unbelievers has been one of those common cases in which both are right and both wrong; Christians being right in their chief assertion, and wrong in standing out for the accuracy of their details, while unbelievers are right in denying that our details are accurate, but wrong in drawing the inference that because certain facts have been inaccurately recorded, therefore certain others never happened at all. Both the errors are natural; it is high time, however, that upon both sides they should be recognised and avoided.

But as regards the demolition of the structure raised in the seventh and eighth chapters of this book, whereinsoever, that is to say, it seems to menace the more vital part of our faith, the ease with which this will effected may perhaps lead the reader to think that I have not fulfilled the promise made in the outset, and have failed to put the best possible case for our opponents. This supposition would be unjust; I have done the very best for them that I could. For it is plain that they can only take one of two positions, namely, either that Christ really died upon the Cross but was never seen alive again afterwards at all, and that the stories of His having been so seen are purely mythical, or, if they admit that He was seen alive after His Crucifixion, they must deny the completeness of the death; in other words, if they are to escape miracle, they must either deny the reappearances or the death.

Now in the commencement of this work I dealt with those who deny that our Lord rose from the dead, and as the exponent of those who take this view I selected Strauss, who is undoubtedly the ablest writer they have. Whether I shewed sufficient reason for thinking that his theory was unsound must remain for the decision of the reader, but I certainly believe that I succeeded in doing so. Perhaps the ablest of all the writers who have treated the facts given us in the Gospels from the Rationalistic point of view, is the author of an anonymous work called The Jesus of History (Williams and Norgate, 1866); but this writer (and it is a characteristic feature of the Rationalistic school to become vague precisely at this very point) leaves us entirely in doubt as to whether he accepts the reappearances of Christ or not, and his treatment of the facts connected both with the Crucifixion and Resurrection is less definite than that of any other part of the life of our Lord. He does not seem to see his own way clearly, and appears to consider that it must for ever remain a matter of doubt whether the Death of Christ or His reappearance is to be rejected.

It is evident that it was most desirable to examine both sets of arguments, i.e., those against the Resurrection, and those against the completeness of the Death; I have therefore mainly drawn the opinions of those who deny the Death from the same pamphlet as that from which I drew the criticisms on Dean Alford's notes. I know of no other English work, indeed, in which whatever can be said against us upon this all-important head has been put forward, and was therefore compelled to draw from this source, or to invent the arguments for our opponents, which would have subjected me to the accusation of stating them in such way as should best suit my own purpose. The reader, however, must now feel that since there can be no other position taken but one or other of the two alluded to above, and since the one taken by Strauss has been shewn to be untenable, there remains nothing but to shew that the other is untenable also, whereupon it will follow that our Saviour did actually die, and did actually shew Himself subsequently alive; and this amounts to a demonstration of the miraculous character of the Resurrection. If, then, this one miracle be established, I think it unnecessary to defend the others, because I cannot think that any will attack them.

But, as has been seen already, Strauss admits that our Lord died upon the Cross, and denies the reality of the reappearances. It is not probable that Strauss would have taken refuge in the hallucination theory if he had felt that there was the remotest chance of successfully denying our Lord's death; for the difficulties of his present position are overwhelming, as was fully pointed out in the second, third, and fourth chapters of this work. I regret, however, to say that I can nowhere find any detailed account of the reasons which have led him to feel so positively about our Lord's Death. Such reasons must undoubtedly be at his command, or he would indisputably have referred the Resurrection to natural causes. Is it possible that he has thought it better to keep them to himself, as proving the Death of our Lord too convincingly? If so, the course which he has adopted is a cruel one.

We must endeavour, however, to dispense with Strauss's assistance, and will proceed to inquire what it is that those who deny the Death of our Lord, call upon us to reject.

I regret to pass so quickly over one great field of evidence which in justice to myself I must allude to, though I cannot dwell upon it, for in the outset I declared that I would confine myself to the historical evidence, and to this only. I refer to spiritual insight; to the testimony borne by the souls of living persons, who from personal experience know that their Redeemer liveth, and that though worms destroy this body, yet in their flesh shall they see God. How many thousands are there in the world at this moment, who have known Christ as a personal friend and comforter, and who can testify to the work which He has wrought upon them! I cannot pass over such testimony as this in silence. I must assign it a foremost place in reviewing the reasons for holding that our hope is not in vain, but I may not dwell upon it, inasmuch as it would carry no weight with those for whom this work is designed, I mean with those to whom this precious experience of Christ has not yet been vouchsafed. Such persons require the external evidence to be made clear to demonstration before they will trust themselves to listen to the voices of hope or fear, and it is of no use appealing to the knowledge and hopes of others without making it clear upon what that knowledge and those hopes are grounded. Nevertheless, I may be allowed to point out that those who deny the Death and Resurrection of our Lord, call upon us to believe that an immense multitude of most truthful and estimable people are no less deceivers of their own selves and others, than Mohammedans, Jews and Buddhists are. How many do we not each of us know to whom Christ is the spiritual meat and drink of their whole lives. Yet our opponents call upon us to ignore all this, and to refer the emotions and elation of soul, which the love of Christ kindles in his true followers, to an inheritance of delusion and blunder. Truly a melancholy outlook.

Again, let a man travel over England, North, South, East, and West, and in his whole journey he shall hardly find a single spot from which he cannot see one or several churches. There is hardly a hamlet which is not also a centre for the celebration of our Redemption by the Death and Resurrection of Christ. Not one of these churches, say the Rationalists, not one of the clergymen who minister therein, not one single village school in all England, but must be regarded as a fountain of error, if not of deliberate falsehood. Look where they may, they cannot escape from the signs of a vital belief in the Resurrection. All these signs, they will tell us, are signs of superstition only; it is superstition which they celebrate and would confirm; they are founded upon fanaticism, or at the best upon sheer delusion; they poison the fountain heads of moral and intellectual well-being, by teaching men to set human experience on the one side, and to refer their conduct to the supposed will of a personal anthropomorphic God who was actually once a baby—who was born of one of his own creatures—and who is now locally and corporeally in Heaven, "of reasonable soul and human flesh subsisting."

Thus do our opponents taunt us, but when we think not only of the present day, but of the nearly two thousand years during which Christianity has flourished, not in England only, but over all Europe, that is to say, over the quarter of the globe which is most civilised, and whose civilisation is in itself proof both of capacity to judge and of having judged rightly—what an awful admission do unbelievers require us to make, when they bid us think that all these ages and countries have gone astray to the imagining of a vain thing. All the self-sacrifice of the holiest men for sixty generations, all the wars that have been waged for the sake of Christ and His truth, all the money spent upon churches, clergy, monasteries and religious education, all the blood of martyrs, all the celibacy of priests and nuns, all the self-denying lives of those who are now ministers of the Gospel—according to the Rationalist, no part of all this devotion to the cause of Christ has had any justifiable base on actual fact. The bare contemplation of such a stupendous misapplication of self-sacrifice and energy, should be enough to prevent any one from ever smiling again to whose mind such a deplorable view was present: we wonder that our opponents do not shrink back appalled from the contemplation of a picture which they must regard as containing so much of sin, impudence and folly; yet it is to the contemplation of such a picture, and to a belief in its truthfulness to nature, that they would invite us; they cannot even see a clergyman without saying to themselves, "There goes one whose trade is the promotion of error; whose whole life is devoted to the upholding of the untrue." To them the sight of people flocking to a church must be as painful as it would be to us to see a congregation of Jews or Mohammedans: they ought to have no happiness in life so long as they believe that the vast majority of their fellow-countrymen are so lamentably deluded; yet they would call on us to join them, and half despise us upon our refusing to do so.

But upon this view also I may not dwell; it would have been easy and I think not unprofitable, had my aim been different, to have drawn an ampler picture of the heart-rending amount of falsehood, stupidity, cruelty and folly which must be referable to a belief in Christianity, if, as our opponents maintain, there is no solid ground for believing it; but my present purpose is to prove that there is such ground, and having said enough to shew that I do not ignore the fields of evidence which lie beyond the purpose of my work, I will return to the Crucifixion and Resurrection.

What, then, let me ask of freethinkers, became of Christ eventually? Several answers may be made to this question, but there is none but the one given in Scripture which will set it at rest. Thus it has been said that Christ survived the Cross, lingered for a few weeks, and in the end succumbed to the injuries which He had sustained. On this there arises the question, did the Apostles know of His death? And if so, were they likely to mistake the reappearance of a dying man, so shattered and weak as He must have been, for the glory of an immortal being? We know that people can idealise a great deal, but they cannot idealise as much as this. The Apostles cannot have known of any death of Christ except His Death upon the Cross, and it is not credible that if He had died from the effects of the Crucifixion the Apostles should not have been aware of it. No one will pretend that they were, so it is needless to discuss this theory further.

It has also been said that our Lord, having seen the effect of His reappearance on the Apostles, considered that further converse with them would only weaken it; and that He may have therefore thought it wiser to withdraw Himself finally from them, and to leave His teaching in their hands, with the certainty that it would never henceforth be lost sight of; but this view is inconsistent with the character which even our adversaries themselves assign to our Saviour. The idea is one which might occur to a theorist sitting in his study, and enlightened by a knowledge of events, but it would not suggest itself to a leader in the heat of action.

Another supposition has been that our Lord on recovering consciousness after He had been left alone in the tomb, or perhaps even before Joseph had gone, may have been unable to realise to Himself the nature of the events that had befallen Him, and may have actually believed that He had been dead, and been miraculously restored to life; that He may yet have felt a natural fear of again falling into the hands of His enemies; and partly from this cause, and partly through awe at the miracle that He supposed had been worked upon Him, have only shewn Himself to His disciples hurriedly, in secret, and on rare occasions, spending the greater part of His time in some one or other of the secret places of resort, in which He had been wont to live apart from the Apostles before the Crucifixion.

I have known it urged that our Lord never said or even thought that He had risen from the dead, but shewed Himself alive secretly and fearfully, and bade His disciples follow Him to Galilee, where He might, and perhaps did, appear more openly, though still rarely and with caution; that the rarity and mystery of the reappearances would add to the impression of a miraculous resurrection which had instantly presented itself to the minds of the Apostles on seeing Christ alive; that this impression alone would prevent them from heeding facts which must have been obvious to any whose minds were not already unhinged by the knowledge that Christ was alive, and by the belief that He had been dead; and that they would be blinded by awe, which awe would be increased by the rarity of the reappearances—a rarity that was in reality due, perhaps to fear, perhaps to self-delusion, perhaps to both, but which was none the less politic for not having been dictated by policy; finally that the report of Christ's having been seen alive reached the Chief Priests (or perhaps Joseph of Arimathæa), and that they determined at all hazards to nip the coming mischief in the bud; that they therefore watched their opportunity, and got rid of so probable a cause of disturbance by the knife of the assassin, or induced Him to depart by threats, which He did not venture to resist.

But if our Lord was secretly assassinated how could it have happened that the body should never have been found, and produced, when the Apostles began declaring publicly that Christ had risen? What could be easier than to bring it forward and settle the whole matter? It cannot be doubted that the body must have been looked for when the Apostles began publishing their story; we saw reason for believing this when we considered the account of the Resurrection given by St. Matthew. Now those that hide can find; and if the enemies of Christ had got rid of Him by foul play, they would know very well where to lay their hands upon that which would be the death blow to Christianity. If then Christ did not go away of His own accord, as feeling that His teaching would be better preserved by His absence, and if He did not die from wounds received upon the Cross, and if He was not assassinated secretly, what remains as the most reasonable view to be taken concerning His disappearance? Surely the one that was taken; the view which commended itself to those who were best able to judge—namely, that He had ascended bodily into Heaven and was sitting at the right hand of God the Father.

Where else could He be?

For that He disappeared, and disappeared finally, within six weeks of the Crucifixion must be considered certain; there is no one who will be bold enough even to hazard a conjecture that the appearance of Christ alluded to by St. Paul, as having been vouchsafed to him some years later, was that of the living Christ, who had chosen upon this one occasion to depart from the seclusion and secrecy which he had maintained hitherto. But if Christ was still living on earth, how was it possible that no human being should have the smallest clue to His whereabouts? If He was dead how is it that no one should have produced the body? Such a mysterious and total disappearance, even in the face of great jeopardy, has never yet been known, and can only be satisfactorily explained by adopting the belief which has prevailed for nearly the last two thousand years, and which will prevail more and more triumphantly so long as the world shall last—the belief that Christ was restored to the glory which He had shared with the Father, as soon as ever He had given sufficient proofs of His being alive to ensure the devotion of His followers.

Before we can reject the supernatural solution of a mystery otherwise inexplicable, we should have some natural explanation which will meet the requirements of the case. A confession of ignorance is not enough here. We are not ignorant; we know that Christ died, inasmuch as we have the testimony of all the four Evangelists to this effect, the testimony of the Apostle Paul, and through him that of all the other Apostles; we have also the certainty that the centurion in charge of the soldiers at the Crucifixion would not have committed so grave a breach of discipline as the delivery of the body to Joseph and Nicodemus, unless he had felt quite sure that life was extinct; and finally we have the testimony of the Church for sixty generations, and that of myriads now living, whose experience assures them that Christ died and rose from the dead; in addition to this tremendous body of evidence we have also the story of the spear wound recorded in a Gospel which even our opponents believe to be from a Johannean source in its later chapters; and though, as has been already stated, this wound cannot be insisted upon as in itself sufficient to prove our Lord's death, yet it must assuredly be allowed its due weight in reviewing the evidence. The unbeliever cannot surely have considered how shallow are all the arguments which he can produce, in comparison with those that make against him. He cannot say that I have not done him justice, and I feel confident that when he reconsiders the matter in that spirit of humility without which he cannot hope to be guided to a true conclusion, he will feel sure that Strauss is right in believing that the death of our Lord cannot be seriously called in question.

But this being so, the reappearances, which we have seen to be established by the collapse of the hallucination theory, must be referred to supernatural or miraculous agency; that is to say, our Lord died and rose again on the third day, according to the Scriptures. Whereon His disappearance some six weeks later must be looked upon very differently from that of any ordinary person. If our Lord could have been shewn to have been a mere man, who had escaped death only by a hair's breadth, but still escaped it, perhaps some one of the theories for His disappearance, or some combination of them, or some other explanation which has not yet been thought of, might be held to be sufficient; but in the case of One who died and rose from the dead, there is no theory which will stand, except the one which it has been reserved for our own lawless and self-seeking times to question. Through the light of the Resurrection the Ascension is clearly seen.

My task is now completed. In an age when Rationalism has become recognised as the only basis upon which faith can rest securely, I have established the Christian faith upon a Rationalistic basis.

I have made no concession to Rationalism which did not place all the vital parts of Christianity in a far stronger position than they were in before, yet I have conceded everything which a sincere Rationalist is likely to desire. I have cleared the ground for reconciliation. It only remains for the two contending parties to come forward and occupy it in peace jointly. May it be mine to see the day when all traces of disagreement have been long obliterated!

To the unbeliever I can say, "Never yet in any work upon the Christian side have your difficulties been so fully and fairly stated; never yet has orthodox disingenuousness been so unsparingly exposed." To the Christian I can say with no less justice, "Never yet have the true reasons for the discrepancies in the Gospels been so put forward as to enable us to look these discrepancies boldly in the face, and to thank God for having graciously allowed them to exist." I do not say this in any spirit of self-glorification. We are children of the hour, and creatures of our surroundings. As it has been given unto us, so will it be required at our hands, and we are at best unprofitable servants. Nevertheless I cannot refrain from expressing my gratitude at having been born in an age when Christianity and Rationalism are not only ceasing to appear antagonistic to one another, but have each become essential to the very existence of the other. May the reader feel this no less strongly than I do, and may he also feel that I have supplied the missing element which could alone cause them to combine. If he asks me what element I allude to, I answer Candour. This is the pilot that has taken us safely into the Fair Haven of universal brotherhood in Christ.

Samuel Butler – A Short Biography

Samuel Butler was born on 4th December 1835 at the village rectory in Langar, Nottinghamshire, to the Rev. Thomas Butler, himself the son of Dr. Samuel Butler, then headmaster of Shrewsbury School and later Bishop of Lichfield.

The young Butler's immediate family created an oppressive home environment that was largely antagonistic (he later chronicled this in 'The Way of All Flesh'). It was said that Thomas Butler, to make up for having been a servile son, became a bullying and overbearing father. The relationship with his mother was better, but not by much.

His education began at home and included frequent beatings, as was all too common at the time. Samuel wrote later that his parents were "brutal and stupid by nature." He later wrote that his father "never liked me, nor I him; from my earliest recollections I can call to mind no time when I did not fear him and dislike him .... I have never passed a day without thinking of him many times over as the man who was sure to be against me."

Under his parents' parochial influence, he was set to follow his father into the priesthood. He was sent to Shrewsbury at the age of twelve, where he did not enjoy the hard life and its routines.

From there, in 1854, he went to St John's College, Cambridge, where he obtained a first in Classics in 1858. The graduate society of St John's was later named the Samuel Butler Room in his honour.

After Cambridge he went to live in an impoverished parish in London 1858–59 as preparation for his ordination into the Anglican clergy; there he uneasily discovered that baptism made no apparent difference to the morals and behaviour of his new peers. He began to question his faith. This experience would later serve as inspiration for his work 'The Fair Haven'. His correspondence with his father about the issue failed to set his mind at rest, inciting instead his father's wrath.

As a result, the young Butler emigrated in September 1859, on the ship Roman Emperor to New Zealand. In common with many British settlers of privileged origins, he wanted to put as much distance as possible between himself and his family. He was determined to change his life and removing himself from the toxic relationship of family seemed the one thing in his control.

He wrote of his arrival and life as a sheep farmer on Mesopotamia Station in 'A First Year in Canterbury Settlement' (1863). After a few years of farming he sold his farm and made a handsome profit. The chief achievement of these years however was not farming or money but his ability to write. In these years were gathered the source materials and the first drafts for much of his masterpiece 'Erewhon'.

'Erewhon' revealed Butler's long interest in Darwin's theories of biological evolution. In 1863, four years after Darwin published 'On the Origin of Species', the editor of The Press, a New Zealand newspaper, published a letter titled 'Darwin among the Machines.' Written by Butler but signed Cellarius (q.v.,) it compares human evolution to machine evolution. Rather startingly it projected that machines would eventually replace man: "In the course of ages we shall find ourselves the inferior race."

But Butler was not a devoted admirer of Darwin. Indeed he spent much time criticising him. In part this was due to Butler's complicated relationship with his own father and grandfather percolating through. He was of the belief that Darwin had used some of the work pioneered by his own father and grandfather and not sufficiently credited it. Butler was an evolutionist, just not of the Darwin kind.

Butler returned to England in 1864, settling in rooms in Clifford's Inn, near Fleet Street, where he would live for the rest of his life.

In 1872, he published his Utopian novel 'Erewhon' anonymously, causing some speculation as to the identity of the author. When Butler revealed himself, 'Erewhon' made him a well-known figure, more because of this speculation than for its literary merits, which remain undisputed.

In 1839 his grandfather Dr Butler had left Samuel property he owned at Whitehall in Shrewsbury on the condition that he survived his own father and his aunt, Dr Butler's daughter Harriet Lloyd. While at Cambridge in 1857 he sold the Whitehall mansion and six acres to his cousin Thomas Bucknall Lloyd, but kept the remaining land surrounding the mansion. His aunt died in 1880 and his father's death in 1886 resolved his financial problems for the last sixteen years of his own life.

There has been some speculation as to why Butler never married. For many years he made regular visits to a woman, Lucie Dumas, where he paid for sex but this seems overshadowed by his intense male friendships, which is reflected in several of his works.

His first significant male friendship was with the young Charles Pauli, whom he met in New Zealand; they both returned to England in 1864 and took neighbouring apartments in Clifford's Inn. Butler now paid Pauli a regular pension to finance his study of the law. This payment continued long after the friendship had cooled. With Pauli's death in 1892, Butler was shocked to learn that Pauli had benefited from mirror arrangements with other men and had died wealthy. He left nothing to Butler.

After 1878, Butler became close friends with Henry Festing Jones. Butler took him on as his literary assistant and travelling companion, at a salary of £200 a year. Jones kept his own lodgings but the two men saw each other daily, working together on music and writing projects during the day, and attending concerts and the theatre in the evenings. They were also frequent visitors to Europe. After Butler's death, Jones edited Butler's notebooks for publication and later published his own biography of Butler in 1919.

Butler was partial to indulging himself, holidaying in Italy every summer and producing his works on the Italian landscape and art. His close interest in the art of the Sacri Monti is reflected in 'Alps and Sanctuaries of Piedmont' and the 'Canton Ticino' (1881) and 'Ex Voto' (1888).

Another significant friendship was with Hans Rudolf Faesch, a Swiss student who stayed with them in London for two years, improving his English, before departing East for Singapore. Butler and Jones both wept when he left on his travels in early 1895. Butler subsequently wrote a very emotional poem, "In Memoriam H. R. F.", which was offered for publication to several leading English magazines. With the

Oscar Wilde trial, and its revelations of homosexual behaviour among the literati now being aired, Butler feared association and hastily withdrew the poem.

He wrote a number of other books, including a moderately successful sequel, 'Erewhon Revisited'. His masterpiece and semi-autobiographical novel 'The Way of All Flesh' did not appear in print until after his death. Butler thought its tone of satirical attack on Victorian morality too contentious and shied away from further potential problems.

Samuel Butler died aged 66 on 18th June 1902 at a nursing home in St John's Wood Road, London. He was cremated at Woking Crematorium, and accounts say his ashes were either dispersed or buried in an unmarked grave.

George Bernard Shaw and E.M. Forster were great admirers of the later works of Samuel Butler, who brought a new form into Victorian literature of utopian/dystopian literature.

Butler belonged to no literary school, and although respected was not the subject of any literary devotion during his lifetime. Although an amateur student of religion and evolution his writings and controversial assertions shut him out from these opposing factions of church and science which played such a large role in late Victorian cultural life: "In those days one was either a religionist or a Darwinian, but he was neither." His later influence on literature came mainly through 'The Way of All Flesh', which Butler began in 1870 and finished after some revisions in 1885 but was unpublished on his explicit wishes until 1903. Despite the decades of its incubation it was still fresh and modern especially in its use of psychoanalytical thought.

Whether it be in his satire and fiction, his studies on the evidences of Christianity, his works on evolutionary thought or in his various other writings, a consistent theme runs through Butler's work, stemming largely from his personal struggle with the stifling of his own nature by his parents, which led him on to seek more general principles of growth, development and purpose. That struggle resulted in a literary career that is still respected to this day.

Samuel Butler – A Concise Bibliography

Darwin among the Machines (1863, largely incorporated into Erewhon)
Lucubratio Ebria (1865)
Erewhon, or Over the Range (1872)
Life and Habit (1878).
Evolution, Old and New; Or, the theories of Buffon, Dr. Erasmus Darwin, and Lamarck, as compared with that of Charles Darwin (1879)
Unconscious Memory (1880)
Alps and Sanctuaries of Piedmont and the Canton Ticino (1881)
Luck or Cunning as the Main Means of Organic Modification? (1887)
Ex Voto; An Account of the Sacro Monte or New Jerusalem at Verallo-Sesia (1888)
The Authoress of the Odyssey (1897)
The Iliad of Homer, Rendered into English Prose (1898)
Shakespeare's Sonnets Reconsidered (1899)
The Odyssey of Homer (1900)

Erewhon Revisited Twenty Years Later: By the Original Discoverer of the Country & His Son (1901)
The Way of All Flesh (1903, also entitled Ernest Pontifex; or, The Way of All Flesh)
God the Known and God the Unknown (1909)
The Note-Books of Samuel Butler (1912)
The Fair Haven (1913, considers inconsistencies between the Gospels)
A First Year in Canterbury Settlement With Other Early Essays (1914)

www.ingramcontent.com/pod-product-compliance
Lightning Source LLC
Chambersburg PA
CBHW060624210326
41520CB00010B/1459